NEW DIRECTIONS IN SATELLITE COMMUNICATIONS

Challenges for North and South

The Artech House Telecom Library

Telecommunications: An Interdisciplinary Text, Leonard Lewin, ed.
Telecommunications in the U.S.: Trends and Policies, Leonard Lewin, ed.
The ITU in a Changing World by George A. Codding, Jr. and Anthony M. Rutkowski
Integrated Services Digital Networks by Anthony M. Rutkowski
The Competition for Markets in International Telecommunications by Ronald S. Eward
Teleconferencing Technology and Applications by Christine H. Olgren and Lorne A. Parker
The Executive Guide to Video Teleconferencing by Ronald J. Bohm and Lee B. Templeton
World-Traded Services: The Challenge for the Eighties by Raymond J. Krommenacker
Microcomputer Tools for Communication Engineering by S.T. Li, J.C. Logan, J.W. Rockway, and D.W.S. Tam
World Atlas of Satellites, Donald M. Jansky, ed.
Communication Satellites in the Geostationary Orbit by Donald M. Jansky and Michel C. Jeruchim
Television Programming Across National Boundaries: The EBU and OIRT Experience by Ernest Eugster
Evaluating Telecommunication Technology in Medicine by David Conrath, Earl Dunn, and Christopher Higgins
Measurements in Optical Fibers and Devices: Theory and Experiments by G. Cancellieri and U. Ravaioli
Fiber Optics Communications, Henry F. Taylor, ed.
Techniques in Data Communications by Ralph Glasgal
The Public Manager's Telephone Book by Paul Daubitz
Proceedings, Conference on Advanced Research in VLSI, 1982, Paul Penfield, Jr., ed.
Proceedings, Conference on Advanced Research in VLSI, 1984, Paul Penfield, Jr., ed.
Advances in Microprogramming, Efrem Mallach and Norman Sondak, eds.
Advances in Computer Communications and Networking, Wesley Chu, ed.
Communication Protocol Modeling, Carl Sunshine, ed.
Advances in Computer System Security, Rein Turn, ed.
Advances in Computer System Security, Vol. II, Rein Turn, ed.
Graphical and Binary Image Processing and Applications, James C. Stoffel, ed.
Principles of Military Communication Systems by Don J. Torrieri
Error-Control Coding and Applications by Djimitri Wiggert

NEW DIRECTIONS IN SATELLITE COMMUNICATIONS

Challenges for North and South

Heather E. Hudson, editor

ARTECH HOUSE, INC.
610 Washington Street
Dedham, MA 02026

International Standard Book Number: 0-89006-162-9
Library of Congress Catalog Card Number: 84-073276

CONTENTS

Acknowledgements

Many people deserve credit for making this conference possible. We greatly appreciated the participation of keynote speaker Richard Colino and luncheon speaker Congressman John Bryant, and Richard Butler, who spoke by teleconference from Washington, D.C., using facilities donated by AT&T. We also benefited greatly from the participation of foreign speakers from Brazil, Canada, India, Japan, and Mexico, as well as the many speakers from U.S. industry, government, and universities. The contribution of the private sector in time and expenses for speakers from industry was appreciated.

I would like to thank Dean Robert Jeffrey of the College of Communication of the University of Texas at Austin for providing financial support which enabled us to bring speakers from developing countries. Support was also received from the Center for Asian Studies and the Department of Electrical Engineering. Assistant Dean Jim Haynes coordinated the preparations and logistics, ably assisted by Julie Williams. Henry Geddes, doctoral student in Radio-TV-Film, helped with numerous planning tasks, and coordinated the efforts of the Radio-TV-Film graduate students who helped with the conference. Graduate students who served as rapporteurs for the sessions were Mavis Bishop, Henry Geddes, Cheryl Harris, Felipe Marcillac, and Lynn York. The recording of the proceedings was carried out under the direction of Joel Fowler, senior lecturer in Radio-TV-Film. Fran Cobb managed the task of producing the final version of this manuscript on the word processor.

Foreword

The conference on "New Directions in Satellite Communications: Challenges for North and South" was held at the College of Communication of the University of Texas at Austin from October 24 to 26, 1984.

The goal of the conference was to examine the relationship of "North" to "South" in terms of satellite applications and policy. The conference examined major initiatives and new satellite services in the industrialized world, and current and projected applications of satellite communications in the developing countries. It addressed requirements for satellite communications in the developing world, as well as policy issues including financing, technology transfer, competition, and access to the geostationary satellite orbit and spectrum.

The keynote speaker was Richard Colino, Director General of INTELSAT, the International Telecommunications Satellite Organization, who spoke on "Closing the Gaps in a Shrinking World: INTELSAT and Rural Satellite Communications." Richard Butler, Secretary General of the International Telecommunication Union, addressed the conference on the second day via audio teleconference from Washington, D.C., on "New Possibilities in Satellite Communications." Congressman John Bryant (D- Texas), member of the House Subcommittee on Telecommunications, spoke at lunch on October 25 on a Congressional perspective on U.S. international telecommunications policy.

Panel topics included:

- New Directions in Satellite Technology and Services
- New Satellite Services: International Implications
- Satellites and the Developing World
- Telecommunications Requirements in Developing Countries
- Future Directions in International Satellite Policy

The Conference brought together executives of the satellite technology and services industries from the United States, Canada, and Japan, officials responsible for satellite planning and policy in several developing countries, and representatives from Canada, Alaska, developing countries, and international aid agencies with experience in using satellite communications for delivery of services to rural and remote areas.

The primary focus was on themes related to the application of satellite technology for development, ranging from experience gained to date, to plans for new systems, to policy issues. The first sessions presented information on technology and services which have been implemented in the industrialized world, but which appear appropriate for developing country needs. The goal of these sessions was to highlight solutions that already exist, but which have not been widely diffused to developing countries.

The second group of sessions surveyed experience gained in using satellite technology for development in northern Canada and Alaska, as well as in developing countries. Then information was presented on the development of satellite communications in India and applications of the INSAT system, and plans for the Mexican domestic satellite (MORELOS) and the Brazilian domestic satellite (BRASILSAT).

Next, some of the unresolved problems which have impeded the diffusion of appropriate satellite technology and other telecommunications facilities to developing countries, including financing, markets, risks and rewards, and training and organization, were discussed.

Finally, on the last day, future directions in international satellite policy were discussed, focusing on the 1985 Space WARC (World Administrative Radio Conference) or ORB 85, where developing countries will seek guarantees of equitable access to the geostationary orbit and spectrum to ensure that they will not be precluded from affordable access to satellites when they are ready to use satellite technology to further their development.

Heather E. Hudson
Austin, February, 1985

OPENING SESSION: INTRODUCTORY REMARKS

NEW DIRECTIONS IN SATELLITE COMMUNICATION: CHALLENGES FOR NORTH AND SOUTH

by

Robert C. Jeffrey, Dean
College of Communication
The University of Texas at Austin

Welcome to the Conference on New Directions in Satellite Communication: Challenges for North and South. The College is pleased to sponsor this conference and to thank our several participants from the United States and abroad for joining us in examining the future direction of satellite communications.

No one need say again how significant has been the developments of satellite communication. We are all aware, I believe, of the revolutionary contraction of our planet made possible by satellite technology. We are aware that such contraction has had, and will continue to have, profound effects on our lives in the areas of international relations, economic development, human rights, health and education. The technology exists to enrich the welfare of people all over the world. It is limited only by the limitations of our intellects and wisdom in using those technologies for proper purposes.

Those who will participate in this conference will be examining over the next two days what we can do to assure that these magnificant technologies will be used for the betterment of mankind. We welcome them and their wisdom and hope that they will make Austin and the College of Communication their home while our guests.

It is my distinct privilege to introduce the keynote speaker for this conference, Mr. Richard Colino, Director General of INTELSAT. INTELSAT, the International Telecommunications Satellite Organization, is headquartered in Washington, D.C. and was created in 1964 when 11 countries signed interim agreements to establish a global commercial communication satellite system. The member countries of INTELSAT now number 109.

INTELSAT was established largely at the initiative of the United States, with the aim of achieving a single global commercial telecommunication satellite system which would provide expanded telecommunications services to all areas of the world without discrimination. Since its first communication services began with the Early Bird Satellite in 1965, INTELSAT has grown until now it consists of 15 satellites in synchronous orbit and, with the global ground segment of 799 communications antennas at 638 earth stations in 157 countries, territories, and dependencies, the system provides 1,274 international communication pathways. In addition to its international communication services, INTELSAT, through individual lease agreements, provides for many developing countries the primary means, and in many cases, the exclusive means, by which they communicate with the rest of the world. Twenty-five countries use the system for their internal domestic satellite services.

At the head of this far-reaching and powerfully-significant organization is Mr. Colino. Mr. Colino has more than 20 years of association with and experience in high-technology telecommunications and broadcasting. He has served as a senior operating executive and a chief executive, with both general management and administrative functions and responsibilities. Before assuming the post as Director General and Chief Executive Officer of INTELSAT, he was president and chief executive officer of a broadcasting and telecommunications consulting firm; president and CEO of a subscription television corporation; and vice president for international operations of the Communications Satellite Corporation where he worked from 1965 to 1979 on international satellite matters. He represented the United States in INTELSAT in many different forums. Richard Colino is the most appropriate person in the world to keynote a conference on new directions in satellite communications. We are fortunate and certainly honored to have him with us to address the topic, "Closing the Gaps in a Shrinking World." Mr. Richard Colino.

Closing the Gaps in a Shrinking World:

INTELSAT and Rural Satellite Communications

Richard R. Colino
Director General
INTELSAT

INTRODUCTION

We live in exciting times! The world we live in has changed more since we were born than it has in the previous century! It has been said that a Moses or an Alexander the Great could have more easily traveled two thousand years into the future to the time of George Washington and Napoleon Bonaparte than Washington or Napoleon could have traveled 200 years into contemporary times. Not so surprising if you just think of the developments which have occurred in your lifetime. We could only speculate how they would react to television, transistor radios, tape decks, nuclear weapons, rockets, computers, robotic manufacturing, or global electronic networks that operate in time increments measured in nanoseconds. In the same context, both the benefits and ill-effects of technology on our health would send them reeling, as they tried to understand things like jet lag and artificial hearts.

Virtually every apsect of living — language, culture, housing, education, work, class structure, and government — has undergone remarkable changes in the last two centuries. If you examined the United States 200 years ago you would find that 85 percent of it — five million people — were engaged in farming; you would also find that trade and financial transactions were largely conducted within the distance of a one-day horseback ride; news from other countries was weeks late in arriving and its impact on the man in the street was of marginal interest when it did occur. Global population was then numbered in the hundreds of millions (rather than some four billion today), and there were only a few dozen countries, rather than almost 200.

Yet there are still parts of the world today where George and Napoleon might feel comfortable in their surroundings. In Bangladesh, they could stil see small villages surrounded by farms with people and oxen at work tilling the fields. On the Congo River in Zaire, one can still see fishermen using techniques hundreds of years old, such as the stringing of wicker baskets about the rapids of the river to catch the leaping fish. These people would also be training their children through stories and fables and instilling in them an appreciation for customs and rituals perhaps thousands of years old!

Then, amid these traditional patterns of living, our time travelers would occasionally encounter a transistor radio, or peculiarly clad teams of workers in a jeep working on an oil drilling rig, or a 747 leaving a trail of white smoke at 35,000 feet. Perhaps in a bigger town they might even see a telephone call box, a community television set — and yes — even a satellite dish. They would begin to wonder where this strange place is — suspended in time halfway in the past and halfway in some mysterious future. They might wonder how some of these peoples had managed to venture into the future, while most of the people seemed to have stood still in the late 18th century. Certainly, if one were to take George and Napoleon to capital cities like Dakar, Brasilia, and Djakarta, and then to rural Senegal, Brazil, and Indonesia, they would have great difficulty understanding the enormous "time gap" that a short "distance gap" of a few miles can make.

Most of you are from developed countries and know that we are living in a complex, shrinking, fast-paced world. Billions of dollars in international electronic funds can be transferred using an INTELSAT V satellite in less than a second. It is routine for airline flights around the world to be booked at a single computer terminal that is instantaneously connected into a global network.

Just as surely, however, in the rural and isolated areas of the developing world, geography, climate, and distance conspire to thwart even the best international efforts at economic and social development. These areas are truly remote in both time and distance. The result has been the so-called "information gap" or "communications gap," on which the UN, the ITU, and other international organizations are concentrating their efforts of late. This "gap" can be measured in many ways, but perhaps most poignantly in the words of the South Pacific islander who said, "To us communications means a boat that arrives once every two weeks." And lack of communication is only one of several problems that need to be solved in order to elevate the level of health care, education, and agricultural development of these countries.

Let me express the communications gap through a few startling statistics.

It is remarkable to me that there are more telephones in Japan than in all of the Third World countries located in the rest of Asia, all of sub-Saharan Africa, and all of South America combined. This seems even more incredible when you consider that Japan has 120 million citizens *versus* the 2.5 billion people who live in Third World countries. In sub-Saharan Africa, one household per 1,000 has a telephone.

Perhaps we should express the problem in terms of financial investment. The amount of capital that would need to be invested to bring telephone density in Third World countries up to U.S. or Canadian standards would be a staggering $7 trillion (U.S.) and this estimate ignores training people to install, maintain, and service such a massive number of telephones.

Where does this leave us? Some would say that the problem of providing a level of communications necessary for the development of the Third World is so massive that the problem is impossible to solve. Should we then write a note to posterity and declare, "We had to forget about this issue — it was an insoluble problem!" Certainly not, and I would like to strike a much more optimistic note. There is simply too much good will, too much energy, and too much knowledge floating around for me to believe that substantial progress cannot be made on this issue. I believe that we can bridge the gap, rather than helplessly watch an ever-widening chasm grow between rich and poor. I say this with confidence as there are several worldwide efforts underway aimed at assembling the necessary parts of the solution into a cohesive plan of action.

I would like to discuss these constituent parts of a solution in the context of three broad categories: namely, technological innovation in the field of rural and remote communications; secondly, positive developments with regard to training and assistance programs; and, thirdly, progress in providing the capital needed to finance communications development.

TECHNICAL INNOVATIONS FOR THE FUTURE

First of all, let's talk about technological innovation. The cost of an installed telephone circuit today, in terms of a line connected to some form of a local exchange switch, ranges between $800 and $1,200 (U.S.). A decade ago, that number would have been $5,000, and well within the next decade that number should drop even more. These cost reductions will result from the application of solid-state electronic technology and the benefits that will be achievable through the application of digital communications techniques. In the satellite world of this past decade, if one had wanted to bring satellite communications to a rural or remote region of the world, it probably would have been necessary to install either an INTELSAT Standard A station costing $10-12 million or an INTELSAT Standard B earth station (costing perhaps $1 million to $1.5 million). Typically, this station would have been installed near the largest city in the region, and would have been linked to the surrounding area by an open wire or HF terrestrial radio telephone link. Not surprisingly, this was expensive. Operationally, the entire system, particularly when connected to terrestrial exchanges, was not necessarily the most reliable or effective.

Today, with the advent of higher-powered and more sensitive satellites, far more cost-effective earth stations are in operation. They use solid-state technology extensively and operate at much lower power levels than previously required. Small earth stations (in the 3-to 5-meter class) can then be installed much nearer to isolated areas.

Indeed, there are already systems available to link these small terminals to a group of, say, 10 to 100 "party-line" telephones, using a technique called "digital concentrators," that can bring high-quality, efficient, long-distnace telephone, and low-speed data service to isolated areas.

However, from a purely economic viewpoint, the revenue generated from such a network can only rarely justify its installation. Only when there is a total commitment to the auxiliary benefits of the communications link, such as one might find for a mining town or construction project, are such systems usually contemplated. So, despite the innovations that have occurred in satellite technology, there is still a long way to go before "direct-to-the-village" two-way voice service is widely available at an economical system cost.

INTELSAT has already taken a first step in this direction, with the introduction of VISTA service in December 1983. VISTA is designed for the thin-route voice service that is required in remote areas, and INTELSAT established a special standard earth station design which permits VISTA users interference-free access to the satellites for only $3,200 per-channel-per-year. With appropriate optimization, our experts believe VISTA terminals can ultimately cost as little as $40,000 to $50,000 and operate on solar power. One can see, however, that for this application the earth station cost is still the dominant factor by far in satellite-based rural service.

What steps has INTELSAT taken to help bring earth station costs more in line with satellite transmission costs? Very recently, INTELSAT introduced a new service called INTELNET. Using tens or hundreds of microterminals costing as little as $2,500 and using antennas as small as 2 feet, INTELNET permits the one-way distribution of printed material, or facsimile, or computer data at 9.6 Kbps. Networks are already operational domestically for Dow Jones and Reuters, and we expect other publications and news agencies to begin international networks shortly.

In short, using these and other examples particularly related to digital satellite communications, the application of solid-state electronics to earth station development and other technological trend lines suggest that major innovations are coming. Technical innovations now foreseen can potentially increase the cost efficiency of rural communications by a factor of ten or more within the next 10 to 15 years. There is great danger here, in terms of promising too much too soon, and giving rise to false expectations. Yet, there is also a counter danger of people doing economic viability studies for rural communications based on yesterday's technology and not anticipating, in terms of planning cycles, the truly impressive technical capabilities that we will see over the next 5 to 10 years.

We at INTELSAT are optimistic about the success of our VISTA and INTELNET services not only for what they mean for rural and remote area communication now, but for the potential they represent for the future.

TECHNICAL ASSISTANCE AND TRAINING

The second area that I wish to discuss is that of technical assistance and training. The telecommunications and satellite industries have a way of focusing on equipment, facilities, and capital investments in ways that at times can be extremely dangerous. There is always a tendency to say, "At first you start with a satellite and go on from there." People who have been involved in the area of rural telecommunications development for several decades know the dangers with this line of thinking. They realize that a satellite is merely a tool that helps to defeat distance and geography, and as a consequence is only a part of the solution.

At the earliest stages, communications planners should start with communications applications and the establishment of priorities among the wide variety of available choices.

They must start with a solid program for developing an overall communications system that responds to users' needs and is complete, in terms of giving the users what they want. The plan needs to provide access to terminal equipment that the users can use. The communications system must have the complete capability to receive or send communications content that is relevant to users' needs (in terms of education, agricultural programs, health and nutrition, and other potential programs). Such programs should take into account the costs of developing the communications content associated with, say, educational radio and television; the cost of not only installing but maintaining equipment; and the tremendous difficulty in obtaining trained personnel that can help improve and maintain a system once it is installed. Real human problems need to be anticipated and addressed, such as, when you do have trained personnel, how can you prevent their departure from the program to pursue job opportunities offering them much higher salaries?

The aspects of human resources and priority of communications content in communications development programs naturally are rather glossed over by telecommunications equipment manufacturers and suppliers of services. However, it has been suggested by some that major communications development programs set aside no more than 10 to 20 percent of their budgets for satellite capacity and procurement of earth stations and long-distance telecommunications equipment. More than 80 percent should be spent on training, maintenance, equipment that is actually used directly in the school or the home, and on software development, in order for a truly comprehensive communications development program to reach its goals. One example of this type of problem has been the Indonesian PALAPA Satellite Program, which has been an outstanding technical success from the viewpoint of satellite communications. Nevertheless, it has also encountered growing pains because the center for educational program production has only recently begun to produce a significant amount of materials — some six years after the program began operation.

This experience highlights an important advantage of the use of the INTELSAT System for domestic communications in developing countries — the fact that capacity can be leased in small quantities that match the commitment of all government ministries to use that capacity, permitting any country to lease only what it needs and no more, and allowing the satellite portion of the domestic communications system to grow economically as the use of applications grows. In fact, we currently lease capacity to some 29 countries, most of which are developing nations. Entities like the U.S. Agency for International Development, the U.S. Technical Training Institute, the ITU's excellent worldwide training program, the INTELSAT Assistance and Development Program (IADP), and like programs sponsored by the Japanese International Cooperation Agency, the DMZ of Germany, and the Canadian foreign assistance agency (CIDA), have all increasingly recognized the importance of improved training programs and better planning and implementation programs for new projects which are now under way.

In addition, the World Communications Year activities in 1983 and the "follow-on" activities associated with the Maitland Commission established pursuant to actions of the 1982 ITU Plenipotentiary Conference in Nairobi and more properly called the Independent Commission for Worldwide Communications Development, have focused, and will continue in the future to focus, important attention on the need for training and technical assistance in the planning of new rural communications programs. The INTELSAT Assistance and Development Program, which was started in 1978, has now provided technical assistance to more than 70 countries. We feel that this program has been of enormous importance in the planning of satellite communications projects in ways that have avoided serious under- or overprovision of facilities and ensured their effective utilization once complete. INTELSAT, of course, has done some strange things under its IADP Program, such as explaining in some considerable technical detail to Nigeria why it was leasing too much capacity from INTELSAT and should be making more effective use of the capacity it had already leased.

In summary, then, I believe that technical assistance, effective overall communications planning for user needs, and training, are key and interrelated elements that will need to be strengthened vigorously over the next decade if the overall goal of rural communications development is to be achieved.

FINANCING COMMUNICATIONS DEVELOPMENT

A final point which I wish to discuss is that of providing financing for communications development in rural and remote areas.

For a number of years, individuals concerned and interested in communications development, including the current Secretary General of the ITU, Mr. Butler, have been, in essence, preaching a sermon — a sermon to which I subscribe. They have said to everyone who would listen that communications development is a key and integral part of any comprehensive attempt at Third World economic development, and that to leave out communications in planning can be every bit as disastrous as ignoring electric power, roads, or schooling. There is increasing evidence that, in terms of the multiplier-effect, investment in telecommunications has an impact on economic growth rates more than almost any other investment. Economic studies consistently show that investing in communications development is fundamental to the overall developmental process. In fact, one of the ITU's studies has concluded that the ratio of benefits to costs of communications services in a typical Third World country was as high as 100 to 1.

A number of studies of the economic impacts of communications development have also concluded that the developmental banking organizations (such as the International Bank for Reconstruction and Development, the Inter-American Development Bank, the Asian Development Bank and others) have devoted too few resources to communications development and, even when resources have been allocated, they have been subject to strict guidelines, under which only marginal improvements can be made. Typically, these lending organizations have devoted four percent or less of their resources to communications-related projects.

The Maitland Commission, in its studies, has tended to conclude that one of the most important contributions would be to increase the capability in the world economic system to provide communications-related investment in developing countries. We at INTELSAT agree. Indeed, we are concerned that national policies which might be designed to stimulate new high-capacity communications systems on high-capacity routes such as the Atlantic Ocean Region, could not only lead to over-investment in unneeded telecommunications facilities, but serve to siphon off monies for capital investment in rural telecommunications services.

INTELSAT has been studying this problem for some time and is deeply concerned. Certainly, we would agree that the problem of providing capital for new investment in rural communications development is a high priority. And INTELSAT, in collabortion with the Maitland Commission, has initiated a serious study concerning the possible creation of a Development Fund. This approach would have INTELSAT serve as a catalyst in working with its members, multilateral and national aid agencies, developmental lending agencies, and others to identify sources of capital financing not only for earth station investment that might be needed for rural satellite communications, but also to finance related terrestrial equipment that is a crucial part of the overall network. INTELSAT will not, however, itself be a major source of direct funding — this would be inappropriate to INTELSAT's operational mission.

INTELSAT feels that this integrated cooperative approach to providing capital is a key step forward. Although this is (in relationship to the global need) a modest program, we feel it is the first major step in this crucial process of finding new and innovative ways to further the cause of Third World communications development. We have always been a leader in the technological aspects of satellite communications, and we are committed to demonstrating that leadership in providing new services, new innovations, new financing concepts, and an ever-stronger commitment to fulfilling our responsibility to the global telecommunications community.

OTHER INTELSAT INITIATIVES IN COMMUNICATIONS DEVELOPMENT

There are, of course, many new and innovative ways to approach the problems of Third World communications development, and INTELSAT is exploring these as well. One of the other parallel initiatives that INTELSAT has recently undertaken, which is of some relevance here, is a program we call Project SHARE, which stands, in English, for Satellites for Health and Rural Education. This is a 16-month-long test and demonstration program, during which INTELSAT will make available free satellite capacity to organizations wishing to demonstrate the beneficial effects of satellite communications in the fields of health and education. Projects will be selected by an International Advisory Council, which INTELSAT has established with the cooperation and assistance of the International Institute of Communications, the co-sponsor of Project SHARE. All Project SHARE tests and demonstrations, of course, must be successfully coordinated with participating signatories in the countries involved.

We are giving priority to projects designed to become operational at the end of the 16-month period; we are, as much as possible, trying not only to demonstrate new and innovative technologies (such as the use of spread-spectrum interactive microterminals) but, in effect, new technology that can be operationally implemented in a cost-effective and meaningful way in the near future. It is our hope that literally millions of people may ultimately share in the benefits of Project SHARE — either in terms of improved health services, newly available educational services, or other applications of which we are not yet aware.

CONCLUSION

In closing, I would like to reassert that while the challenges are great, the possibilities of significant new improvements are beginning to fall into place. Technological innovation in the next decade should push us toward operational equipment that in some cases may be up to ten times more cost efficient. The importance of training and technical assistance has been recognized by a growing number of institutions and entities, and expanded capabilities are emerging around the world. The world's development agencies and financial communities have become better educated in the need for increasing the capital available for communications development.

Thus, we truly do live in an exciting world. The gaps in communications — even in a shrinking world — are enormous but not unbridgeable. We know that improved communications mean development, and development is essential for world peace and stability. INTELSAT has fulfilled its role through innovations in services and technology, and by maintaining a policy of keeping rates for use of its satellites as low as possible. Assuming INTELSAT is allowed to continue to exist as a viable global cooperative with a mandate to bridge the gap between the East and West and North and South, and to maintain global interconnectivity, then much can and will be done in the decades ahead. At this critical time, many crucial issues in the communications field hang in the balance. Let us hope that, when the final measure is taken, INTELSAT will be encouraged to complete its many initiatives for the future. With positive answers, INTELSAT, working with its global partners, should indeed, in time, ultimately banish remoteness of peoples and space on our planet.

PANEL 1: NEW DIRECTIONS IN SATELLITE TECHNOLOGY AND SERVICES

The purpose of the first session was to provide examples of technologies and services developed in the industrialized world that are thus already available and are likely to be appropriate for developing countries with little or no modification. The speakers were representatives of a major U.S. provider of data and voice communications which also provides international services, the vice president of a very large Japanese company operating throughout the industrialized and developing world, and the president of a small U.S. manufacturer of satellite equipment for voice and data communications, particularly for remote locations.

The first speaker, Mr. Ryoji Tamura, is Executive Vice President of NEC America, Inc. (Nippon Electric Corporation). In 1983, he was transferred to NEC America, Inc. from the parent NEC Corporation in Japan where he had worked for 27 years after graduating from Hokkaido University. Mr. Tamura worked mainly in the development of microwave and satellite communications equipment. Before joining NEC America, Inc., he was general manager of the microwave and satellite communication division at NEC Corporation. Mr. Tamura's presentation described satellite communications technology developed by NEC specifically for rural areas. He described the parameters and design of a satellite system optimized for rural telecommunications, modulation/coding techniques, and the configuration of thin route networks using small earth stations.

The second speaker, Mr. Ronald Stowe, is currently Vice President for Government and Commercial Affairs of Satellite Business Systems, headquartered in McLean, Virginia. SBS now has four satellites in orbit, and emphasizes primarily business resources for international services, and for relations with Congress and the Executive Branch of the U.S. government for SBS. He was a member of the U.S. delegations to the 1983 Regional Administrative Radio Conference (RARC) on Broadcast Satellites and to the 1979 World Administrative Radio Conference (WARC). Before joining SBS, Mr. Stowe was the Assistant Legal Advisor for United Nations Affairs at the State Department. Mr. Stowe's paper outlines many of the recent innovations in voice and data services via satellite. He pointed out, however, that proponents of satellite technology have a responsibility not only to encourage the development of the technology, "but also to ensure that the institutional and political framework which grows up around it will enhance and not undermine its usefulness."

The third speaker, Mr. Ray Sensney, is Chairman of the Board and Chief Executive Officer of Dalsat, headquartered in Plano, Texas. Mr. Sensney has had a long career in the satellite industry. In the early 1960s he was a member of the launch team of SYNCOM 1, the first geosynchronous satellite. In the mid-1970s, he designed the original WESTAR system for Western Union, and later designed the PBS network system while at Rockwell International. In 1978, Mr. Sensney founded Dalsat. In his presentation, Mr. Sensney described many of the products and services offered by Dalsat which have applicability for the developing world, including transportable earth stations, a "portable telephone booth," earth stations for relaying data from offshore oil rigs and remote drilling sites, equipment for compressed video teleconferencing, and an agricultural marketing network.

SATELLITE COMMUNICATIONS TECHNOLOGY FOR RURAL AREAS

by

Ryoji Tamura

Executive Vice President, NEC America, Inc.

INTRODUCTION

It is not easy to provide reliable telephone, telex, facsimile and data communications even in cities and built-up urban areas. And in a rural zone, the long distances, difficult terrain, and relatively small number of subscribers make it all the more difficult to provide economical communications. Historically, rural communications have suffered in relation to urban communications.

A rural zone generally consists of scattered settlements, villages, and small towns, where the zone exhibits one or more of the following characteristics:

a) scarcity of primary power, or uncoordinated, scattered, power generation;

b) scarcity of locally available qualified technical personnel;

c) topographical conditions which are obstacles to the construction of conventional lines and transmission systems, e.g. lake, desert, snow-covered or mountainous areas;

d) in some zones, tropical, semitropical or other severe climatic conditions that make critical demands on the life and maintenance of equipment;

e) economic constraints on amortizing investments and rendering service profitable, due to high costs of construction and maintenance, if they are to be borne by the rural zone alone, with a possibly quite restricted economic base.

However, satellite communications can play a significant role as a first step towards the development of the telecommunications infrastructure in a rural area. Because a satellite can cover a wide area, overcoming distance and natural obstacles, high-quality wideband communication is possible between earth stations within the satellite coverage area. Since an earth station can be installed at almost any time and any place where a communications need exists, satellite communications can efficiently and flexibly promote telecommunications development in rural areas. In addition, as they do not require repeater stations, as is the case of terrestrial links, satellite systems require less maintenance and are thus suitable for rural areas where skilled maintenance engineers are unavailable.

Satellite systems offer high-quality wideband satellite channels which can directly connect two or more points, without using existing telecommunications facilities.

Satellite communication is flexible enough to provide diversified services ranging from conventional telephone service to more modern enhanced communications services by simply replacing parts of existing terminal equipment according to communications needs. Therefore, no surplus investment in communications equipment is required. The minimum equipment, to meet initial communications needs only, may be installed and extra equipment can easily be added later if and when the need arises.

GENERAL REQUIREMENTS FOR THE RURAL SATELLITE COMMUNICATIONS SYSTEM

The network and the earth station design effort for "Rural Satellite Communications" would be focused on the following major requirements.

(1) Frequency Band to be Used

The rural satellite system has to cover equatorial areas where rainfall is generally heavy and which, anyway, have a long rainy season. Communications have to be maintained in this environment, and it is generally important to secure emergency communications even during disasters caused by heavy rainfall. The use of frequency bands which are less sensitive to rain attenuation is generally preferable for these areas.

(2) Earth Station Equipment

Earth stations to be installed in rural areas should employ state-of-the-art technologies. They should be compact and simple in structure and consume the minimum of electrical power.

These earth stations should incorporate the following features, among many others:

a) The design needs to minimize power requirements to allow the use of solar cells or other alternative power sources and to minimize the use of conventional power where it is available. In order to meet this requirement, at least transmitting power should be less than several watts.

b) The design should seek to reduce the size of earth station antennas in order to keep the capital costs of the earth station to a minimum and to allow easy transportation and installation in rural locations without special construction equipment. In addition, antennas located in rural areas should be of a fixed type without an automatic or a manual tracking feature.

c) Earth station equipment housed in a simple shelter should be capable of continuous operation and able to withstand severe environmental conditions.

(3) Multiple Access Techniques

The rural satellite system should employ simple state-of-the-art techniques proven in field operation rather than complex, unproven techniques. From this viewpoint, the single channel per carrier (SCPC) technique, which uses FM or PSK modulation, seems to be most appropriate for rural satellite systems, as it is efficient for both analogue and digital transmission, and can efficiently serve a large number of earth stations.

Among channel assignment techniques, the demand assignment multiple access (DAMA) technique with a master control station seems best for rural satellite systems. This technique allows efficient utilization of satellite channels and can do away with or simplify earth stations switching equipment.

(4) Communications Satellites

Satellites for rural communications systems will require high performance, e.g., high figure of merit (G/T), high equivalent isotropically radiated power (e.i.r.p.) and high gain, to offset economic and performance limitations of the rural earth stations.

The satellites should be exclusively used for local communications purposes. Therefore satellites shared by a number of countries need not radiate global beams at all. A single spot beam can be assigned to rural areas where earth stations share the same transponders.

In order to meet specifications (e.g., high G/T and e.i.r.p.) the satellites should have high-gain antennas as far as satellite attitude stablization and satellite capacity allow, so that sharp beams can be focused on their assigned areas only.

Table 1 shows the relationship between satellite antenna gains and beam coverage on equatorial surfaces around a subsatellite position.

Table 1: RELATIONSHIP BETWEEN SATELLITE ANTENNA GAIN AND BEAM COVERAGE

ANTENNA GAIN (in dBi)	BEAM COVERAGE (in km sq)
18	GLOBAL
26	5,000
29	3,000
34	1,600
38	1,000

Assuming that rural satellite systems cover a global zone from 45°N to 45°S, the total land portion in this zone can be covered by 80 "15° x 15°" segments, and the most of the sea portion by 16 "30° x 30°" segments.

The earth surface covered by a "15° x 15°" segment amounts to approximately 1,600 km square; and that covered by "30° x 30°" segment, approximately 3,000 km square. In order to serve the whole of the above zone, a total of about 96 transponders with 29 to 34 dBi antenna gain, or 4 satellites each with 24 transponders will be required.

APPLICABLE CURRENT SATELLITE COMMUNICATION TECHNOLOGY FOR RURAL AREAS

(1) Communication Satellites

As discussed in the previous section, from a technical view point, the most important factor in applying satellite communication systems to rural areas would be how to realize a high gain antenna economically for a rural communication satellite to offset a low e.i.r.p and/or a low G/T performances of rural earth stations.

This technique, however, has been developed and put into a practical use in the INTELSAT system (1) as well as the other DOMSAT systems.(2)

INTELSAT spacecraft antennas have evolved from providing a single coverage global beam, as on INTELSAT III, to multiple coverages, as on INTELSAT V to achieve multiple frequency reuse through spatial and polarization isolation between various coverage beams by employing a multiple shaped beam antenna technology.

All technical efforts of the antenna design were concentrated on obtaining more zone beams with closer spacing, wider frequency bandwidth and less circuit loss, keeping adequate spatial and polarization isolation between geographically separated coverage areas, while maintaining a homogeneous illumination over every dominant coverage zone.

These efforts resulted in making it possible to mount a 2.4m offset reflector at 4 GHz transmit band fed by a feed array of some eighty horns on INTELSAT V or a 3.2m offset reflector with a feed array of one hundred and several tens of elements on INTELSAT VI, whereas previously only a 1.2m parabolic reflector with a single feed horn was used on INTELSAT IV.

For the satellites to be exclusively used for rural communication purposes, a single spot beam interconnected with one or two transponders would be enough to illuminate one coverage zone on the earth surface, where earth stations share the same transponders to communicate with each other, in, say, a "1600 km square" or a "3000 km square" coverage area as mentioned before. The system design, in this case, should also take into account the actual traffic demands, geographic conditions and so on, of all the rural areas. It is possible that neither closer spacing nor wider frequency bandwidth would be required for the antenna beams, which have a common frequency band and the same polarization sense, because isolation between the closely separated beams is easily attainable by using a "frequency spacing" technique.

A 2.4m reflector fed by a single feed unit can illuminate over 1000 km square of equatorial surface around a subsatellite position and a 3.2m one will cover about 800 km square of the same. It is clear that a multiple shaped beam antenna for a rural communication satellite with an antenna gain of 29 dBi to 34 dBi and segmented by multiple beams each of which covers a 3000 km square to a 1600 km square earth surface, can be easily produced by applying developed and field proven technology using an offset reflector antenna fed by horn array feed. The antenna diameter can be the same diameter as on an INTELSAT satellite or a smaller diameter with a fewer number of feed elements than the INTELSAT satellite.

Concerning the spacecraft itself, no other technical difficulties are foreseen in developing a communication satellite for rural communications.

(2) Modulation/Coding Techniques

Of the various kinds of multiple access technologies available, single channel per carrier (SCPC) system provided with a demand assignment multiple access (DAMA) function is one of the most appropriate systems for a rural satellite communication network, where several hundred subscribers' terminals, each of which handles limited long distance multi-destination traffic, are scattered over an extensive area and interconnected via a satellite transponder.

In order to keep a good communication quality even at a lower receiving level at the earth station, various modulation/coding techniques have been investigated and examined during the past decade.

Thanks to this development effort, many modulation/coding techniques are applicable to a rural satellite communication system, and some of them are already put into practical use in the INTELSAT and DOMSAT systems though prudent choice of the modulation/coding technique for the initial planned rural system will be necessary.

A companded FM-SCPC (CFM-SCPC), which was initially adopted for the "STAR" system (3) -- a root of the "SPADE" system -- and now employed in so many DOMSAT systems is one of the most suitable techniques for voice signal transmission, while an ADPCM PSK-SCPC (ADPCM-SCPC), which is now under study by the CCITT may be another suitable technique for voice and data transmission. The former can be characterized by a simple and an economical system, whilst the latter can be characterized by providing high quality transmission performance but is a complex system.

By applying either of the modulation/coding techniques mentioned above to a rural satellite communication system, the antenna diameter of an earth station can be reduced, without any degradation to the system performance to about 0.6 compared with a current standard system employing a 64 kbps PCM PSK-SCPC (PCM-SCPC).

Table 2 indicates the technical comparison of these modulation/coding schemes.

(3) An Earth Station

By adopting the current available technology as discussed above, that is,

a) a rural communication satellite characterized with a high antenna gain of 29 dBi to a 34 dBi, and

b) a modulation/coding technique, using either CFM-SCPC or an ADPCM-SCPC, the antenna diameter and transmit output power of earth stations for rural communication systems can be reduced to two to three meters and several watts or less, respectively.

Figure 1 shows a photograph of a typical small earth station currently produced by NEC Corporation, Japan, for a DOMSAT system.

Fig. 1 A TYPICAL SMALL EARTH STATION

Fig. 2 NEC SMALL EARTH STATIONS

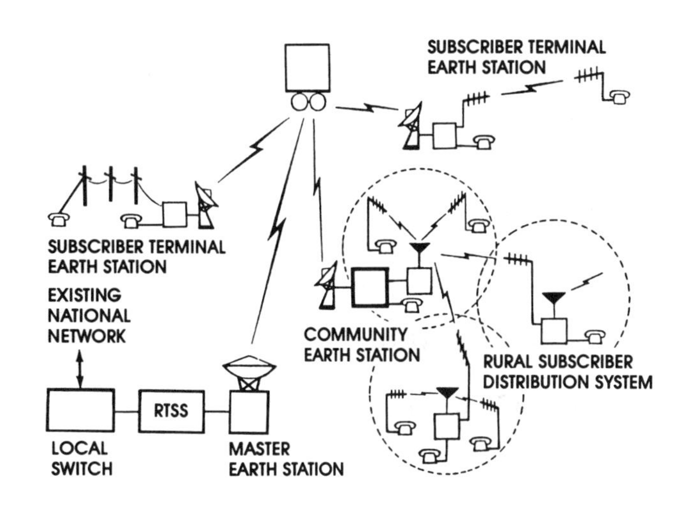

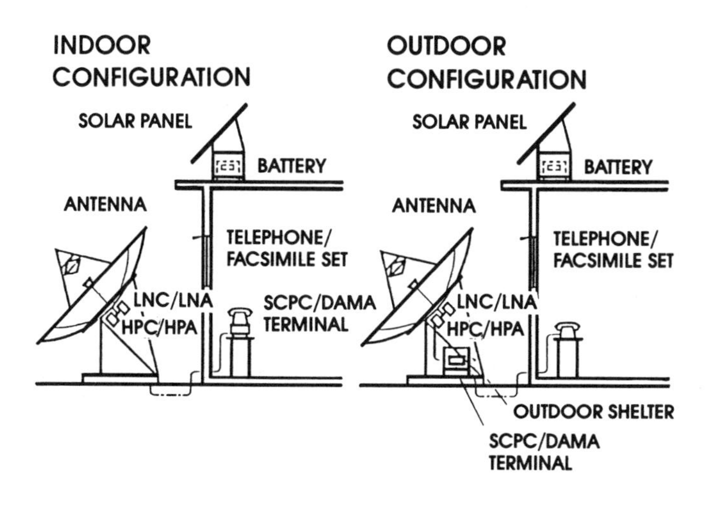

Fig. 3 TYPICAL LAYOUT OF NEC SMALL EARTH STATION

Fig. 4 OUTPUT POWER AND POWER CONSUMPTION OF A SUBSCRIBER TERMINAL EARTH STATION

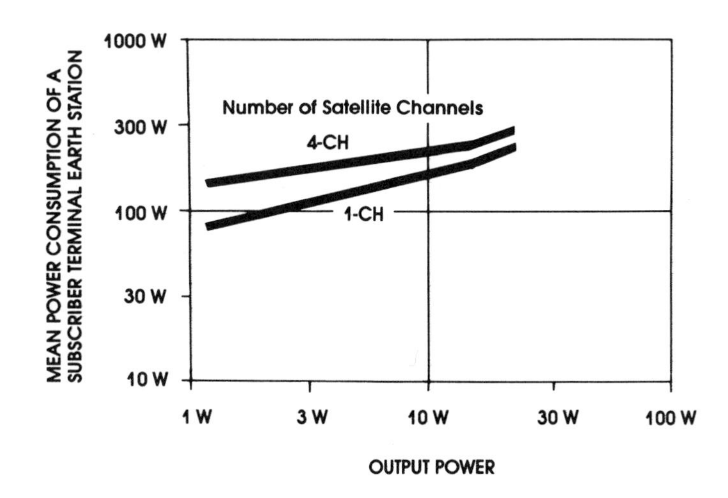

Table 2: TECHNICAL COMPARISON OF TYPICAL MODULATION/CODING SCHEMES

SCHEME	PCM PSK	ADPCM PSK	COMPANDED FM
STATEMENT			
TRANSMISSION BIT RATE/BAND	64 kbps	33.3 kbps	0.3 to 3.4 kHz
VOICE QUALITY	EXCELLENT	EXCELLENT	GOOD
INFORMATION BIT RATE	56 kbps	32 kbps	4.8 kbps
CIRCUIT CONFIGURATION	COMPLEX	COMPLEX	SIMPLE
C/N AT THRESHOLD LEVEL (approx.)	59 dB/Hz	54.5 dB/Hz	54 dB/Hz

The NEC earth stations are divided into three different types according to their functions, usage and scale as illustrated in Figure 2.

They are a subscriber terminal earth station, a community earth station and a master earth station.

The subscriber terminal earth station may be installed at either a public booth or a private homestead in remote areas, and furnishes subscribers facilities, such as telephone sets, facsimilies, teletypewriters, etc., as required.

The community earth station may be installed at a communication node in either a rural district or a customer premises in a remote area to interconnect with a rural subscriber distribution system, PABX, and so on.

The master earth station may be installed at any place, if it is within a satellite beam coverage, and acts as the regional center for rural subscriber exchange lines (satellite circuits) in the dominant area and also interconnects with existing national telephone networks (trunk circuits) through toll switching equipment.

In order to provide a switching function to a rural satellite communication local network, all earth stations involved are provided with a central controlled DAMA equipment, so called "Remote Telephone Satellite Switching System" (RTSS) developed by NEC Corporation.

The main features and advantages of the RTSS are as follows:

a) Single hop connection between any two remote earth stations can be accomplished automatically and double hop connection problems within the same beam coverage can be avoided.

b) No additional channel, other than the communication channel, is required for either subscriber terminal or community earth stations, because each SCPC channel unit is designed to be commonly used for signalling and communication exchanges.

c) All necessary functions which are required for DAMA operation, such as supervisory, control, automatic accounting and report, etc., are centralized at the redundant master stations and then, unattended reliable operation can be performed at other remote earth stations.

A typical layout of the NEC's subscriber terminal earth station with one telephone/facsimile set is illustrated in Figure 3 and the power consumption for up to four channels is shown in Figure 4.

Where AC commercial power is unavailable, it is necessary to use a power supply system which consumes less energy and requires no maintenance specialists.

A solar cell system, consisting of solar cell batteries, is an ideal power supply source because it requires little maintenance and consumes no fuel.

Since solar cells systems, generating up to 300 watts, are more economical than any other power generating system now, a 3.3 meter-3 watt subscriber terminal earth station equipped with one to four CFM or ADPCM SCPC satellite channels will be the most suitable for earth stations which can only be installed where commercial power is unavailable.

Although a small earth station operating in the KU band (14/12 GHz) has been described, C-band (6/4 GHz) version is also available.

CONCLUSION

By applying such satellite communication technologies as described above to a rural satellite communication system, each transponder with a 40 MHz bandwidth can service 500 to 1000 earth stations with a traffic intensity of about one erlang per station.

Table 3: NEC SMALL EARTH STATION

MAJOR UNITS	PARAMETERS
FREQUENCY RANGE	Ku BAND
ANTENNA (in diameter)	2.1/3.3/4.5/6.4 meters
LOW NOISE FET AMPLIFIER	180/220/350 k
POWER AMPLIFIER	
FET	1/1.5/3/6 Watts
TWT	15/40/100 Watts
SCPC TERMINAL	
MODULATION	C-FM/ADPCM/PCM
CH CAPACITY	1 to 14 CH

Satellite communication technologies accumulated during the past twenty years, based on numerous and remarkable research and development programs most of which have already been used on the INTELSAT system and DOMSAT systems, now seem to be approaching a level where they can be put into practical use for a rural communication system.

REFERENCES

1. C. C. Han, et al. "A General Beam Shaping Technique-Multiple-Feed Offset Reflector Antenna System." Montreal, AIAA/CASI April 1976, 76-249.

2. E. T. Pfund. "Regional Satellite Systems for the Late 1980s." Orlando, AIAA March 1984, 84-697.

3. M. Morita et al. "STAR System." NEC Research & Development No. 8, October 1966.

NEW VOICE AND DATA SERVICES: DOMESTIC AND INTERNATIONAL MARKETS

by

Ronald F. Stowe

Vice President, Government and Commercial Affairs,

Satellite Business Systems

During the past twenty-seven years since the launch of the first man-made satellite into orbit, nations around the world have launched thousands of objects into outer space. These spacecraft have performed functions ranging from atmospheric testing to meteorological monitoring, from civilian natural resources sensing to deep space research.

Some of the most conspicuous programs for the use of satellites during the past decade has been the development and wide-spread use of space stations for the relay of telecommunications. These programs have been initiated by private organizations, by national governments and by multilateral international organizations. They have so far ranged from the tiny Telstar to the huge INTELSAT VI, which is scheduled for launch in 1986, with a capacity of approximately 30,000 equivalent voice circuits (plus at least two television channels). Telstar relayed telephone, data, video and facsimile signals among more than 90 countries on four continents.

At the present time, a number of countries are operating their own domestic satellite systems. Other countries are utilizing regional systems for a combination of domestic and regional services, and over 160 countries are utilizing the multi-national systems of INTELSAT and INTERSPUTNIK for communications links around the world. In addition, an increasing number of countries are using capacity provided by INTELSAT for domestic as well as international services. In brief, over 170 countries, representing the full spectrum of economic and technological development, are now utilizing satellite systems to help meet their telecommunications needs.

Because of this growing reliance on satellite technology, we are meeting here in Austin to compare our experiences and insights. In turn, our goals are to improve our mutual understanding and if possible to make some progress toward a broader international consensus on the key policy issues. In this context, however, it is particularly important to keep a realistic perspective on the actual nature of the problems we confront. Our essential challenge is to establish efficient and progressive policies and institutional arrangements which will help us all meet our domestic and our international telecommunications needs.

Satellite telecommunications policy is only one part of that challenge. In some countries and regions it has major practical importance; in others it may or may not be a very significant factor in solving real telecommunications needs.

Satellites in fact represent only a relatively small percentage of the total communications facilities which are being used around the world. For most countries, including the United States, the twisted pairs of copper wire, coaxial cables, microwave, UHF, cellular and other terrestrial radio facilities form both the traditional and still the predominant means of transmission. The introduction of optical fiber cables in the coming decade, both on land and under the sea, will reinforce that fact, and will make available tremendous amounts of new capacity for both domestic and international services.

These facts in no way minimize the extremely important contributions which satellite technology is making and will in the future continue to make to us all. They do, however, support the argument that we should not heap the full burden of solving domestic or North/South economic, technological or social development challenges onto the back of satellite telecommunications policy just because it is new, dramatic, filled with impressive potential, and not yet tainted by failure. It is one new tool which can be used in our efforts to encourage economic development; it is a fantastically exciting technology; but it is not a panacea.

If we try to make it carry too heavy a load beyond its obvious functions, satellite telecommunications policy can asily become so encumbered and so constrained that we will risk losing many of the advantages which now hold such great promise in the telecommunications service arena. As experts, afficionados or general enthusiasts about satellite telecommunications, we have a special responsibility not only to encourage the development of this new technology, but also to ensure that the institutional and political framework which grows up around it will enhance and not undermine its usefulness.

In fact, of course, satellite technology is off to a racing start in providing telecommunications services. Although satellites were initially used primarily for long-distance trunking of analog telephone signals, the industry is now moving rapidly into the age of digital transmissions which combine voice, data, video and facsimile into a single bitstream. This technical innovation has introduced a level of efficiency and flexibility which is having quite a dramatic impact on the costs and appeal of such services. These innovations have been applied as much to the improvement in cost and performance of existing telephone services as they have to the introduction of new applications such as high-speed data and videoconferencing.

By using a combination of digital format and techniques such as voice compression activity, it is now possible, for example, to send up to four voice channels over the same capacity which in the past could handle only one. By using a digital format, users now also have the option of maintaining a single system for both voice and data, often using the off-peak periods at night for low-cost data distribution, rather than maintaining two separate networks around the clock.

Improvements in the costs and performance of voice services, prosaic as they may sound, are extremely important both to large businessess and to individual subscribers, both in this country and abroad. Public switched telephony is currently a multibillion dollar a year business in this country alone (and that is just for the long-distance portion of the call). In addition, as competition and lower prices have been introduced in the United States during the last several years, the marketplace has demonstrated a surprising and very substantial elasticity in demand. In other words, individuals, businesses, and other organizations want to and will use the telephone more than they have in the past if the costs come down and services are diversified.

Deregulation and competition among United States carriers have had a positive effect not only on our domestic telecommunications services, but also on the international services which people in this country have available. Although the introduction of satellite technology was itself not a significant motivating factor in the domestic policy changes which lead to deregulation and increased competition, satellite technology has in fact played a very significant role in accelerating the actual availability of new efficiencies and services here and abroad. There are a number of good examples.

Within the United States, the introduction in 1980 of satellites designed to provide flexible and sophisticated high-speed data circuits has led to the expansion of multiple systems capable of providing comparable services; it has also been a significant factor in accelerating the schedule for introducing more flexible and higher capacity terrestrial systems by established carriers which had no competitive motivation to do so earlier.

Although the analogy is not exact, it is also of interest to compare the impact of potential competition in the international services areas as well. Although INTELSAT had clearly been considering the introduction of more flexible and attractive services for several years, it is also reasonably clear that the proposed introduction of competing systems had a substantial positive impact on the introduction of new and attractive international satellite services by INTELSAT earlier this year.

Those services, referred to as International Business Services (IBS), promise to bring substantial new benefits not only in the contested North Atlantic region but to all other interested countries, North and South as well. The prospect of increased international satellite competition was clearly one of the key factors in the timing and nature of the innovations now offered by INTELSAT. Because INTELSAT services virtually the entire world, countries in South America, Africa and Asia will be able to benefit from these changes, not simply the countries in the North Atlantic.

And what are the new benefits which are now being realized?

Although Mr. Colino has certainly addressed INTELSAT's plans in some detail, it may be worthwhile to hear from a domestic carrier's point-of-view what looks most promising on the international scene. As noted above, one of the most exciting developments in this area has been the introduction of the IBS range of services. From our perspective there are two key elements to these new services. First, IBS offers high-speed services on a part-time or occasional use basis, rather than requiring customers to buy such circuits on a full-time basis, whether they are needed all the time or not. These part-time options can lower the cost and expand the interested market, encouraging a much broader range of customers to use services such as high-speed data and videoconferencing.

Second, IBS permits users to transmit signals directly to INTELSAT spacecraft high over the Atlantic, and soon over the Pacific oceans. Although both options will still be available, it will no longer be necessary first to carry all international signals to remote national gateways before sending them overseas. It will soon be feasible to send signals originating in New York, a few blocks or a few miles, to any one of several international earth stations in Manhattan or the surrounding area, rather than 200 miles out to Etam, West Virginia. Signals from Chicago will go overseas directly from Chicago; those from Austin will go directly from Austin. Eventually it will also be feasible in many cases to send signals from the west coast of the United States directly to Europe through only one satellite hop. In each case, services will be less expensive and more flexible than ever before.

In domestic systems comparable advantages are already available. The recent introduction of systems using the new

12 and 14 GHz (Ku) bands has made it feasible to position high-powered earth stations in the middle of urban centers, where the services are most often needed. In the past, when such domestic systems were limited to using the 4 and 6 GHz (C) bands, it was often essential to place the earth stations in more remote or shielded areas to protect the signals from harmful interference by terrestrial microwave services which use the same frequencies.

And how are all these new facilities and techniques being used? Large companies use high-speed satellite links to move large volumes of data among business locations in very short periods of time. A 300 megabit computer disc, containing the records of a day's business transactions, can, for example, be transmitted from coast to coast either by terrestrial facilities or by satellite. By terrestrial facilities it would take approximately 13.5 hours, and so it is not done. A courier service which had carriers duplicate tapes is used instead, with all the risks which that entails. By satellite facilities at 1.5 megabits per second (Mbps), the same amount of data can be sent in less than 30 minutes; at 3.0 Mbps it would take only 7 minutes.

This increased speed has many benefits. It frees the computers on both ends for other uses. It permits load sharing among computer facilities in remote locations and in different time zones. It permits small field offices to tap into the master computers at headquarters, reducing pressures to duplicate expensive facilities just in order to serve remote locations.

Satellite technology may well provide the most flexible and efficient medium for national and international distribution of electronic mail. Satellites are already being used by the American Bar Association, the Association of Western Hospitals and many other geographically dispersed organizations to provide continuing education on a nation-wide basis. Satellites have for several years been used to distribute Spanish language television programs from Mexico into the United States. Satellites have been used in Alaska to help local paramedics provide first level medical care to remote villages.

Universities throughout the country have already begun to link their campuses by satellite, not only to exchange and share the lecturing skills of their professions, but also to establish a much broader network for the exchange of research materials and data bases. Satellites are also now being used by oil exploration companies to link seismic crews in remote locations directly with analytical computers in Tulsa, Houston and Dallas, permitting real-time analysis of the seismic results.

In addition to these types of telecommunications links, satellites are apparently on the verge of introducing the long-waited direct-to-home television broadcasting service. This service could provide an extremely valuable function in both the entertainment and education areas. It would be immediately accessible from any point in the satellite's footprint, regardless of terrain and distance from major urban centers. Major obstacles still remain in the path to successful implementation of this service; the costs of equipment installation and maintenance must be perceived as competitive and affordable, and adequate quantity and quality of programming must be found if subscribers can be expected to sign up in large numbers. These problems are being worked on, however, and it is presumably a matter of time, not a question of whether they can be resolved.

In brief, the continued growth in satellite technology both in the United States and in other countries, for both domestic and international telecommunications, is beginning to generate very exciting and very practical benefits for users, large and small. There will be a further surge in the number of communications satellites which are launched in the next several years, and that added capacity is likely to have two very beneficial results for customers. First, it is likely to continue to bring prices down, both because the available capacity is growing and because additional competition is being introduced. Second, it is likely to generate a great amount of innovation and flexibility in the types of services that are made available. We will find new, more interesting and more productive ways to use that capacity, and we are all, in developed and developing countries alike, going to be the beneficiaries. Let us be certain that we encourage and not stifle the emergence of those benefits.

INNOVATIONS IN SATELLITE TECHNOLOGY APPROPRIATE FOR THE DEVELOPING WORLD

by

Ray Sensney

Chairman of the Board, Dalsat

INTRODUCTION

Dalsat Inc. is a satellite communications systems engineering company located in Plano, just north of Dallas, Texas. Dalsat is a total systems company that takes the responsibility for end-to-end performance of a communications network. Although our principal expertise is in satellite communications, we may be required to use microwave, fiber optics, VHF radio or open air laser communications to satisfy our customers' requirements. Most of our products are interface equipment although when required, or when customer sponsored, we will build main line hardware.

We take responsibility for all phases of design, engineering and construction of end-to-end networks. We have the technical capability of assembly and integration of products, whether our own or some other manufacturer's, into a fully integrated system.

We are involved in the legal aspects of coordinating frequency plans with cross-country microwave. Since the frequencies are shared, there is the potential for mutual interference and, therefore, interference measurements are an integral part of site selection. We also do the technical calculations that are required as part of the FCC license application.

Dalsat provides completely integrated systems including mechanical engineering and installation and civil engineering and construction. We also engineer and design prime power systems including generators, uninterruptable power systems, power line protection and regulation.

EQUIPMENT

Transportable Earth Stations

One of our products is a transportable video uplink which is used by all the major U.S. television networks and was used for the Superbowl this past year. Other Dalsat stations were used for both recent political conventions.

Dalsat has built over 70 transportable earth stations in this or a similar configuration, and we currently have 18 in the manufacturing process.

Fixed Earth Stations

We manufacture products in support of our systems design. As a smaller company with limited resources, we choose to design and manufacture radio products which are transparent to channel dependent hardware. Other products, such as power amplifiers, are readily available from established suppliers.

Non Standard INTELSAT Stations

Dalsat built six meter INTELSAT earth stations for use in Saudi Arabia. The networks used a PABX controlled demand assignment system. Thus the computer in the PABX can also be used to search for a vacant channel on the satellite system.

Other Products

We manufacture products in support of our systems design. As a smaller company with limited resources, we choose to design and manufacture radio products which are transparent to channel dependent hardware. Other products, such as power amplifiers, are readily available from established suppliers.

Up-Converter

The DSA-6 up converter is similar to devices that have been used for years in support of INTELSAT requirements.

With the increasing use of TDMA, up converters with high frequency stability and phase noise performance are required.

Down Converter

The DSA-4 is the companion unit to the DSA-6 and has particular application to high speed data networks. We have a DS 8000 microprocessor control system that is used to control radio and switching equipment.

Antenna

The secret to our success in the transportable earth station market is our proprietary design for a fold-up antenna system. This antenna system must be capable of folding to less than 8 feet for road transportation, but when unfolded for use, must maintain a service accuracy of better than .030 of an inch. We are designing a new earth station that can transmit television from anywhere on the earth, and can be flown by commercial airline.

With the accuracy and repeatablity, this antenna system qualifies for the very short wavelengths of Ku-band satellite systems. This particular station was used by NBC television as part of their study to determine the feasibility of using Ku-band satellite for network television.

APPLICATIONS

Dalsat builds earth stations for many applications that include voice, data and television.

Portable Telephone Booth

Some earth stations are used at remote locations, for example, oil and gas sites in the mountains of Wyoming. The nearest telephone is fifty miles away.

Emergency Restoral

AT&T uses transportable earth stations for emergency restoral of voice communications in case of fire, flood and other disasters.

High Speed Data

We are involved with high speed data transmission with some of our customers. Often, we find it necessary to go to our customers' premises for the installation of data terminal equipment.

Remote Low Speed Data Transmission

Dalsat supplies a small antenna with dedicated electronics and batteries for use on pipelines and off-shore oil rigs. This is a 9600 bit per second transmit-receive data terminal.

Teleconferencing

Oil companies with a shortage of geologists use teleconferencing to transmit satellite data from a remote site to geologists in conference rooms at headquarters. Video compression is used to reduce the bandwidth required.

Ku-band Studies

Last year under contract to NBC television, Dalsat conducted a comprehensive evaluation of the effects of rain on Ku-band television. This study is recognized as the most comprehensive data base available today on the subject. With NBC's permission, the results of this study will be released later this year.

Public Broadcasting Service

The Public Broadcasting Service was the first U.S. television network to distribute their program material via satellite. For the past five years, Dalsat has enjoyed a service contract to implement five or six stations per year.

Cellular Radio

In 1981, Dalsat designed and built a satellite interconnect system for VHF radio. This system has twelve nodes and is the forerunner of today's cellular radio concept. The person in the car talked through the VHF repeater to the satellite earth station and via satellite to the head office. This design could be used for ship-to-shore or off-shore rigs.

CONCLUSION

The purpose of this presentation has been to stimulate your thinking by showing you a range of equipment and applications available today. I hope you have also been able to use your imagination to determine how these technologies and services could be applied in developing country contexts.

DISCUSSION

QUESTION:

Will SBS enter a market which requires smaller earth station terminals? Mr. Colino indicated that INTELSAT would pursue a similar market with its INTELSAT service.

MR. STOWE:

SBS is now in the process of expanding not only our private network for large customers, which was our original market, but also our public switched network. We are offering a digital service which is targeted for residential and small to moderate sized businesses, carrying both voice and data, which is comparable to both MCI and Sprint. Our short run strategy to gain revenue and market share is to increase our services to the individual, small business and medium-sized business user in both voice and data. At the same time, but for different reasons, we have decided to offer smaller antennas because we have found that our customers really do not need all the features of very large antennas. We have currently approximately 120 antennas in service of the 5 meter to 7 meter diameter. Within the next two years, we expect to see 600 to 1000 smaller sized, two way antennas; however, the technology is not available at this time. Small antennas for receive only service are now available, but they cannot serve the needs of customers needing several megabits of interactive digital transmission.

QUESTION:

What size do you foresee for the smaller antennas you are discussing, and will these new antennas be capable of either two-degree or one-and-a-half degree spacing?

MR. STOWE:

Currently, no supplier offers a 2 to 3.5 meter antenna which meets the two degree spacing requirement at affordable, attractive prices. There is no point, from the user's point of view, in purchasing an antenna which does not at least meet the two degree spacing requirement. We foresee using a 2.5 to 3.0 meter antenna for commercial customers with two-way communications requirements. For receive only needs, .6 to 1.0 meter antennas are now in use. There is a cost trade-off to be considered in using a smaller antenna; the smaller the size of the antenna, the more sophisticated other portions of the equipment must be to discriminate the signal effectively.

QUESTION:

Where do you see the technology for off-set fed antennas for earth stations?

MR. STOWE:

SBS is now using an off-set feed on our .6 meter antennas that we now serve from our satellite. The satellites that we put up originally had 20 watts of transmit power; we are now moving into 40 watts of transmit power and spot beam capability for increased flexibility. We are a good example of upgrading our equipment to meet new customer demands and technical options. The off-set feed feature on some of our smallest antennas makes it possible to provide smaller diameter antennas.

QUESTION:

Could you tell us the diameter of a 9.6 KBPS capacity K-band terminal? Will this terminal be able to meet CCIR requirements? If you made a C-band version, what would it cost?

MR. SENSNEY:

A 9.6 KBPS, K-band terminal has a 2 meter antenna, constructed so that it can be shifted in a single piece. The system is prime focal fit, with an offset support system, so there is only one interfering member, in effect. The FCC Rules and Regulations (Part 25) specify that the off-axis performance mandated is applied to the orbital arc only; therefore, the left-right plane must be proper while the up-down plane is free to vary. By holding the prime focus feed mechanism in the up-down plane so that irregularities in pattern occur in this plane, one can achieve the required two-degree spacing.

The two degree spacing requirement is generally directed toward television transmission. The spacing regulation is one way to establish parameters for adjacent satellite interference or energy density without specifying interference levels directly. The 9.6 KBPS channel or any FM SCPC system, requires a very low power energy density, so that the two degree spacing rule does not really apply to voice grade transmission in the same manner as it does for television transmission. The two meter spinning antenna is consistent with the FCC's more stringent standards, albeit through a strict intepretation of the rules.

In looking at the cost of comparable C-band terminals, one has to examine the trade-off between recurring and non-recurring cost. Some industries, such as the oil industry for example, have a desperate need for voice communication as well as extreme physical barriers to communication. In this case, much of the service costs be regarded as a function of the transponder allocation and not so much related to the antenna itself.

QUESTION:

How do you maintain a focus on a satellite from an offshore oil rig? Do the rigs rock on the water, or are they somewhat stabilized?

MR. SENSNEY:

There are three answers to the questions because there are basically three different kinds of rigs: those that are stabilized, those that are semi-submersible (work in shallow water), and those that are true floaters like the Ocean Ranger. The objective is to set up the antenna so that the twisting motion of the rig does not exceed the dB gain width of the antenna as installed. The fact is, however, that there are no really satisfactory solutions to the rig motion. Some companies make antenna-mount devices which are gyro-controlled for the true floaters or for shipboards, but these devises are not particularly successful.

(Comment from the audience): Unless you use the INMARSAT System.

MR. SENSNEY:

Yes, but if you use the INMARSAT system then you must use the INMARSAT satellite, which is, in effect, DC power limited, consequently limiting the number of available working channels. It costs a good deal every time someone talks for a minute over one of the INMARSAT channels.

QUESTION:

Considering the state of the art in satellite technology, how will the technology translate into the needs of the developing countries?

MR. SENSNEY:

My goal today has been to impart ideas. While I have been to several developing countries, I cannot understand nor anticipate all the needs of developing countries. I have learned from experience that our customers come to us with an idea. They want us to help them develop that idea into something achievable, a a communications solution that can be realized. I believe that all three of us here today, including Mr. Colino, have provided background concepts and a picture of the state-of-the-art, and I hope the developing countries can take back those ideas, turn them into requirements, and bring back those ideas to the satellite industry as a whole.

MR. STOWE:

Let me give you a couple of satellite applications (for developing country communications). Basic telephone communication is the most important thing to the developing world, just as it is to the United States. There are all sorts of stages of development, but basically a country would want to start, if it can, with an efficient, affordable ability to call anywhere in the country, and then expand service to reach the rest of the world. Many of the world's rural areas are served by radio, which has quality problems and capacity limitations. Mexico is a good example of a country which had a stated national policy to provide an adequate telephone network to connect the villages, providing for safety and good commerce.

Satellites are not necessarily the answer to all developing country communications. Where terrain is difficult, satellites can provide a very nice answer.

Another important satellite application is video transmission. Television services are used for education and to further other cultural and economic goals.

On the other hand, we have many domestic customers who are very anxious for developing countries to have improved telecommunications. Many of the big domestic customers are multinationals, which, for a variety of reasons, are looking more and more to increasing their manufacturing and assembly functions in other countries (other than the United States). A growing and important factor in these business decisions to transfer operating functions is the availability of good telecommunications services within the host country.

MR. TAMURA:

I will approach the question from a technological viewpoint. We can provide a three to eight meter earth station and transponder as I described before using a proven technology. This implies that a single spot beam antenna can serve up to 1000 square kilometers of surface area. Naturally, using a spot beam antenna, we can reduce the cost of the earth station, and we need some compromise. What I propose today is to combine 49dB antennas and 34 dB antennas, which can serve 3000 x 3000 square kilometers in surface area. A large antenna would be useful in this case. In other situations, a 34 dB antenna could serve 1600 x 1600 square kilometers surface area.

TELECONFERENCE WITH RICHARD BUTLER

Mr. Richard Butler is the Secretary General of the International Telecommunication Union, headquartered in Geneva, Switzerland. Established in 1865 as the International Telegraph Union, the ITU took its present name in 1932, and since 1947 has been a specialized agency of the United Nations. Its membership now includes 158 nations. The purposes of the ITU are to maintain and extend international cooperation for the improvement and rational use of telecommunications of all kinds, to promote the development of technical facilities and their most efficient operation with a view to improving the efficiency of telecommuincations services, increasing their usefulness, and making them, so far as possible, generally available to the public, and to harmonize the action of nations in the attainment of those ends. The ITU works through conferences which provide directives to the permanent organs of the Union which include the General Secretariat, the International Frequency Registration Board (IFRB), the International Radio Consultative Committee (CCIR) and the International Telegraph and Telephone Consultative Committee (CCITT). The ITU also provides technical assistance in the form of expert advice and training through its Technical Cooperation Department.

Mr. Butler was unable to come to Austin, but agreed to address the conference and to answer questions via audio teleconference from Washington, DC. His presentation described the possibilities of using appropriately designed satellite systems to meet the needs of developing countries, and outlined current ITU activities in assisting developing countries to obtain basic telecommunications infrastructure.

NEW POSSIBILITIES IN SATELLITE COMMUNICATIONS

by

Richard Butler

Secretary General

International Telecommunication Union

Ladies and Gentlemen:

It is indeed a privilege to join in your international symposium by teleconference, and to present to such a distinguished gathering as yours a brief overview of the Union's work in the area of satellite communications.

At the outset I wish to draw attention to the fact that one of the basic objectives of the Union is to coordinate the efforts of nations with a view to harmonizing the development of telecommunications facilities, notably those using space techniques.

Over the past two decades, many dramatic advances have been made in satellite communication technology.

In tune with this, the Union has also updated its regulations, agreements and standards to meet the needs of the new medium. As you know, these standards, we call them recommendations, are drawn up by the International

Consultative Committees of the Union while the regulations which often include basic technical parameters are drawn up by a number of World and Regional Administrative Radio Conferences of the Union starting from the 1963 Space Communications Conference.

In this context I should only like to add that next year in August-September, will be held the first session of the World Administrative Radio Conference relating to the use of the geostationary-satellite orbit and the planning of space services using it. A conference preparatory meeting of the CCIR in July/August this year has established the technical information to assist this session in 1985.

The first session will provide an overview of the international arrangements for the various space services, and, *inter alia*, will then decide on which services should be subject to different planning and/or coordination approaches in an international context.

The second session of the conference will be held in 1988. The purpose of orbit conferences is to assure arrangements which guarantee in practice for all countries equitable access to the orbit and the frequency bands allocated to space services.

In this regard we see the consequences of the success, especially the new economies in the use of space technology. Costs continue their downward trends in a remarkable way, and hence the greater interest of many new countries in wishing to choose the appropriate satellite system specifications for their particular requirements -- individually or collectively.

I wish now to turn to the work of the Union in regard to the practical applications of satellite communication for development.

Right from the advent of communication satellites, it has been recognized that communicating via satellite has a number of unique benefits such as distance - as well as terrain - independent costs and quality of performance, network flexibilities of the highest order including point to multi-point and multipoint to point transmission capabilities, the convenience of use of transportable or mobile terminals as well as relative ease in the setting up of links.

It was the first mentioned advantage that led to the early use of the medium for transoceanic communications in the global network. However, with the dramatic drop in satellite communication space and ground segment costs, the medium has progressively found application in regional, sub-regional and domestic situations as well. Many examples come to mind even in the developing world such as the PALAPA, ARABSAT and INSAT systems. It has been the Union Secretariat's privilege to be associated in one way or another with the evolution of some of these latter systems.

A significant trend has been the complexity inversion between space and ground stations. Whereas to start with, satellites were relatively simple with the result that large earth stations had to be used, it is present practice to have satellites of greater power and higher sensitivity with the possibility of concomitant reductions in ground station size and complexity. The full advantages of these inversions are only now being appreciated. Far too often space has been regarded as "add on" to say a national network or to serve large provincial areas within national networks. Far too few have seen space as a prospective and now economic competitor in the available selections of transmission means, be they the interesting varieties of cable, radio and space systems to service large or collections of smaller user communities.

Today, I should like to refer to one of the studies that the Union has carried out concerning the possible new applications of satellite communications. This study was concerned with the use of high power satellites, or transponders of satellites, for domestic rural and remote area communications. The system recommended by the study is one that would be available to developing countries to provide intra-country connections over which thin route communications could be established. With a satellite designed for high sensitivity, transfer gain and power, with a steerable spot or area beams, it would be possible to use ground stations with 3m and less antennas and fully solid-state equipment including power supplies suitable for operation in unelectrified rural areas.

For many developing countries a judicious blend of satellite communication technology with traditional terrestrial solutions now appears to offer a unique opportunity for rapid expansion of their national networks in directions they prefer and to reach remote and rural areas sooner than would have otherwise been possible.

In spite of the substantial cost reductions that have resulted over the years, the use of satellite technology does, however, call for significant investment in resources not only financial but also technical. Because of this underlying requirement it appears essential that in the use of satellite communication there ought to be greater sharing of resources and facilities wherever possible. It is in such sharing that great opportunities for international cooperation exist.

The sharing of resources is indeed a well tried and successful approach. There are the examples of INTELSAT, INTERSPUTNIK, INMARSAT and PALAPA satellite systems. Not only are economies of scale achieved by such sharing, but also technology transfer and manpower development are facilitated. Furthermore, the sharing need not be confined to North-South cooperation with appropriate technology transfer, but can include South-South collaboration as well.

Two examples of such collaborative initiatives in which the ITU has played and continues to play a catalytic and supportive role are worth mentioning.

The first relates to the development of a regional satellite communication system for the development of Africa. In this project and in response to a Resolution by Ministers, the Union is working as the lead agency in close cooperation with African regional organizations and UNESCO (for the media) with a view to advice on the selection of an optimal mix of terrestrial and satellite based facilities which could best serve the telecommunication requirements within the African continent.

The second relates to the use of satellite communications in the development of national and regional networks for the South Pacific island states which are themselves widely separated and are each made up of a number of islands by no means close to one another. For this sub-region, bilateral as well as ITU studies are proceeding through the South Pacific Bureau for Economic Cooperation with a view to establishing overall plans for the evolution of the appropriate mix of terrestrial and satellite-based systems concerned. Unlike in the case of Africa (where one of the desires is for a regional satellite to provide the required space segment capacity for many countries), offers have been forthcoming in the South Pacific on the potential availability of transponder capacity on a second generation Australian domestic satellite (AUSSAT).

These are but examples of how cooperative efforts of the type now well established for global communication systems can be extended to provide the means to respond to sub-regional and domestic requirements, particularly to areas denied the benefits of service in the past.

There exist now excellent opportunities for international collaboration, in the sharing of satellite facilities tailored to thin route needs in particular, to bring at a low cost services to those communities unable to be served by technologies of the past.

However, so varied and great are the unfulfilled needs, that clearly there is much scope for further international collaboration in this field. Naturally, there are important policy and cooperation questions which arise in the sharing of international resources and facilities for providing regional and domestic services. Nevertheless, given the past record of the international community in regard to such sharing, it can be expected that the appropriate responses will eventually be forthcoming.

Before concluding, I should like to touch on two other aspects.

The first relates to an aspect to which the Union attaches great importance. This concerns the development of manpower resources that would stimulate and sustain the endogenous growth of satellite and related communication technology. Examples of action taken in this regard are those connected with the establishment or development (with ITU support) of research/training centres in India and Brazil as well as the PACSATNET projects in Indonesia. There are many other ITU activities with more or less the same objective. I refer also to various seminars and meetings which the Union organizes with a view to introducing and familiarizing technical personnel of developing countries with the practical applications of satellite communication technology.

The second task relates to the dissemination of information on satellite communications, especially to the developing world. In this context, there are many publications of relevance to the subject issued by the Union. The handbooks prepared by the International Consultative Committees are particular examples.

I have in this brief presentation devoted attention mainly to the developmental aspects of the Union's work in the area of satellite communications.

This is not in any way to detract from the fundamental role the Union plays both in its own right and as the UN specialized agency for telecommunications in the development of satellite communications standards and regulations which enable and facilitate the harmonious growth of this unique medium. Nor is it because I underestimate the challenges of standardization and regulation.

It is my belief, nevertheless, that given the excellent spirit of international cooperation for which the Union has a long and distinguished record of nearly one hundred and twenty years these challenges, including the guaranteeing of equitable access to all countries to the geostationary orbit, will be met in due measure.

However, in the sharing of resources and facilities, despite the many promising starts and, indeed, some significant achievements, we still seem to have a long way to go. Considering the great potential that the use and effective deployment and integration of satellite communication technology into our global network offers the establishment of adequate communication infrastructures so vital to socio-economic development, the challenge of resource and facility sharing by North-North and with and by South-South cooperation must be given a priority.

I am confident that, with mutual understanding and goodwill, a way can be found to spread more effectively, and in a judicious manner, the benefits of satellite communications to all.

DISCUSSION

QUESTION:

When will the report be available which updates ITU members on the experience of the application of Articles 11 and 13 from WARC 79?

MR. BUTLER:

The report is now available. It will most likely be a subject of discussion during preparatory sessions prior to the Space WARC.

QUESTION:

Have these been substantial contributions to the ITU's Special Voluntary Fund?

MR. BUTLER:

There have been some contributions. For instance, a training center is now underway in Zimbabwe; Finland is supporting and planning a network development in Sri Lanka; multilateral cooperation is being channelled to support studies which could not be undertaken by one nation alone.

QUESTION:

What activities will the ITU undertake to assist developing countries to prepare for the Space WARC?

MR. BUTLER:

Pending funding negotiations, the ITU expects to announce a pre Space WARC conference meeting in developing regions by the end of 1984. For the Asian-Pacific region the seminar will probably be in May. For Africa, it will be at the end of April or early May, and for Latin America, in March.

QUESTION:

Could the ITU serve as an institutional framework to support the GLODOM satellite project?

MR BUTLER:

In my opinion, the ITU's future role will be one of facilitating and advising. To date, the ITU has served this function with respect to regional proposals in Africa and among subregional groups such as the Andean nations.

QUESTION:

How can technology transfer to developing countries be facilitated?

MR. BUTLER:

The South should not be considered as a homogeneous grouping with identical needs and assets. Countries such as India, for example, have already got the capability to produce for their own technological requirements. The key issues in technology transfer are the extent to which technologies have been adapted to the local environment and the conditions of sale. INTELSAT has been instrumental in promoting the use of satellite services. However, little has been done in the area of technology transfer to induce local production. Countries such as India, Brazil, South Korea, and Singapore have made moves in this direction, largely on their own initiative. This is a policy issue which needs to be taken up at the national level.

QUESTION:

Can you comment on why the costs of ITU documents are so high? They are effectively inaccessible to some organizations that need them.

MR. BUTLER:

The rates for printed material will be reviewed at the policy levels of the Administrative Council itself. Generally, prices are coming down.

QUESTION:

Would you comment on the importance of thin route narrow-band satellite communications versus the traditional wide-band services for networks in developing countries?

MR. BUTLER:

Thin-route satellite communication makes it possible for developing countries to have basic two-way communication. One of the objectives of the African communications decade was to bring within walking distance a central common user facility. ITU studies have shown that investments in telecommunications for the rural sector result in economic and social benefits.

QUESTION:

What progress is being made to expand reliable telecommunications in the South Pacific regions?

MR. BUTLER:

Regarding the South Pacific region, leaders within that region have concluded that there ought to be a coordinated program for communications development. They also concluded that their future inter-island needs depend upon an appropriate satellite facility. Studies are needed to assess those facilities and their financial requirements. In the meantime, national networks will continue to expand.

PANEL 2: NEW SATELLITE SERVICES: INTERNATIONAL IMPLICATIONS

The purpose of this session was to continue the examination of new technologies and services developed in the industrialized world which appear to be suitable for meeting developing country needs, and to begin the examination of technologies and services specifically designed for developing country applications.

The first speaker, Mr. Raul Rodriguez, is an attorney for PanAmSat, the Pan American Satellite Corporation, a company which proposes to provide telecommunications and video services for Latin America. Before joining PanAmSat, Mr. Rodriguez worked for the National Telecommunications and Information Administration (NTIA). PanAmSat has applied to the Federal Communications Commission for authority to build, launch, and operate a satellite system designed to meet the telecommunications needs of Spanish-speaking nations of the western hemisphere. Mr. Rodriguez' paper outlined the technical design and services proposed to be offered by PanAmSat.

The second speaker, Dr. Bruce Lusignan, is Director of Stanford University's Satellite Communications Planning Center, and Associate Professor of Electrical Engineering at Stanford. Dr. Lusignan has been involved for many years in planning and designing satellites systems and equipment for developing regions. Dr. Lusignan was involved with the ATS-1 and ATS-6 satellite projects, and assisted the State of Alaska in designing its small stations for village communications. Dr. Lusignan has also consulted for the governments of several developing countries in planning rural telephone networks and thin route satellite services. His presentation outlined many of the types of equipment which have already been developed in the U.S. and could be suitable for developing country use, and some of the impediments to transferring such appropriately designed and relatively inexpensive technology.

The third speaker, Mr. George Davies, is Director of Space Applications for the Canadian Department of Communications. Mr. Davies has been involved for the past decade with applications of satellite technology for education, social services, business, and community development. The Department of Communications has assisted potential users of satellite technology to gain experience through participation in projects and experiments on the Communications Technology Satellite (CTS) and the ANIK-B satellite. The conditions in Canada's remote north are similar to those in many developing regions. Mr. Davies paper outlined the lessons from the Canadian experience in satellite applications that may be relevant for developing countries.

PAN AMERICAN SATELLITE CORPORATION: NEW OPPORTUNITIES FOR LATIN AMERICAN TELECOMMUNICATIONS DEVELOPMENT

by

Raul R. Rodriguez

Attorney, Pan American Satellite Corporation

Good morning. I am honored to be here -- a conference on North-South space communications -- since I represent the first company ever to propose to initiate a commercial satellite system to serve the domestic communications needs of South America and North-South video traffic. I am Raul Rodriguez, attorney for Pan American Satellite Corporation, which has applied to the FCC for authority to build, launch and operate the first satellite system specifically designed to meet the telecommunications needs of the Spanish-speaking countries of the Western Hemisphere.

Despite what Mr. Colino said yesterday regarding satellite systems apart from INTELSAT, PanAmSat is not a scheme to "cream-skim" or to compete with INTELSAT for high density route traffic. PanAmSat's objective simply is to provide Latin American nations with access to their own bulk satellite capacity which they can use to meet their individual countries' communications needs and to facilitate Spanish-language video and audio transmissions among all Spanish-speaking people -- two objectives which INTELSAT has largely ignored.

Briefly, let me describe what PanAmSat proposes to do. PanAmSat will offer transponders for sale or lease on a non-common carrier basis to both U.S. and foreign entities, including networks, independent program producers, cable systems, television news and wire services and governments. These transponders will be marketed for both domestic communications services within individual countries in South and Central America and the Caribbean Basin, as well as Hemisphere-wide video and audio transmission services.

To help you better understand PanAmSat's configuration, refer to the diagrams on the following pages. Figure 1 shows PanAmSat's domestic service, comprised of 24 transponders dedicated exclusively to domestic service within Central and South America and the Caribbean, operating in the "C-band." These 24 transponders will be configured in three subcontinental beams and one movable beam, which are shown in Figure 1.

One spot beam covers the Caribbean and Central American region. A West Beam covers the Andes region. A South Beam provides coverage of Argentina and Chile. An additional moving spot beam, capable of serving any region in Latin America, will be used to meet additional coverage where demand requires it.

The international coverage (Figure 2) is provided through an additional continental beam with uplink and downlink capability both within South America and to and from the United States, with a small spot beam covering the Iberian Penisula. The remaining 12 transponders thus are intended for regional services, although their use will be limited to video and audio transmission. These 12 transponders will use frequencies uniquely planned to avoid regional technical interference, yet maximize the efficiencies of this limited resource. PanAmSat thus has been planned with state-of-the-art technology to be flexible and capable of meeting the unique telecommunications requirements of Latin American countries.

PanAmSat will join together all of the Hispanic nations of the world, as well as the United States, thus making possible increased cultural, political and economic ties in our Hemisphere. PanAmSat's pursuit of these goals is fully demonstrated by its reservation of transponder capacity for use without charge by the Organizacion de la Television Iberoamericana and by universities throughout the hemisphere.

PanAmSat, however, will provide more than just service flexibilities and new service opportunities. The system will make optimal use of its power and technical configuration to maximize its efficiencies, resulting in very economical space and earth segment costs. While PanAmSat users will be able to access the satellite with 5 meter earth stations costing between $5,000 and $10,000, a comparable quality video signal using INTELSAT satellites would require a substantially larger earth station -- in the magnitude of 10-15 meters -- costing 20 to 50 times as much ($100,000 for a 10 meter antenna and $250,000 for a 15 meter antenna). When you consider each of these in multiples of tens and hundreds -- keeping in mind the numbers of earth stations required for a viable domestic or regional network -- PanAmSat's distinct economic advantages are readily apparent.

PanAmSat's space segment costs also are significantly different from those of INTELSAT. We calculate that costs to end users will be approximately $700,000 per channel year. Compare that figure with INTELSAT's charges of $1.31 million for two-video-channel transponders and $2.28 million for one-video-channel transponders. And, those figures do not represent Signatory mark-ups at both ends.

Cost comparisons, however, are not the only measure of PanAmSat's greater efficiencies. PanAmSat is the only satellite which all of Latin America needs to point to for international Spanish-language television and audio program exchange. PanAmSat thus eliminates the difficulties of dealing with multiple satellite sources and the tremendously increased cost and inconvenience this causes for earth segment plans.

One cannot overlook, however, PanAmSat's ability to provide Latin American countries with access to state-of-the-art technology which heretofore has been available primarily to the most highly industrialized countries. Countries purchasing transponders for domestic use will for the first time have access to their own space segments without having to depend on international boards to decide how much capacity will be made available, for what use, under which priority and at what price. Transponder owners and lessees will also exercise a direct participation in the collective ownership, not just a fractional allocation based on usage figures. A more direct participation in the use of the space and earth segments will facilitate a greater degree of transfer of technology and systems know-how which will enable user-owners to understand better the intricacies of satellite technology and its applications.

In sum, PanAmSat offers:

-low-cost space segment;

-long term non-preemptible leases or sales;

-inexpensive earth station access;

-single satellite "connectivity" to the Spanish-speaking nations of the world;

-highly flexible domestic and international service capability; and

-greater opportunity for North-South transfer of technology and systems operation know-how.

Given the benefits to be derived by Latin American nations and the United States by the creation of a sub-regional communication satellite system dedicated completely to providing new services to a woefully underserved part of the world, we fail to understand the delay in moving forward with the necessary authorizations. The issue of alternative international satellite systems was first raised in March of 1983. Literally thousands of person hours have been spent debating the pros and cons of alternative systems. Five applicants have gone through rounds of administrative procedures, including applications, petitions to deny, replies, allegations and counter allegations. Several Washington private consultants have spent months researching and writing on this topic, not to mention the thousands of hours spent by lawyers and lobbyists.

Congress has also looked into this matter. No less than four or five hearings have focused on this issue -- in the House and Senate. A Senior Inter-agency Group made up of very high level Government officials from 14 different federal agencies studied the matter for more than nine months, and according to press accounts, has recommended to the President that alternatives to INTELSAT, if properly defined in scope, were indeed in the "national interest."

Few issues in the field of telecommunications have come under such scrutiny by so many. And, with the exception of those with a vested interest in the existing regime -- which until recently had failed to provide adequate service to Latin America -- there appears to be unanimity among the participants to this debate that alternatives to INTELSAT, if defined in scope so as not to disrupt the established international public switched networks, would increase the flow of information, make possible mutually beneficial trade opportunities, augment our understanding of each other's cultures, and promote the use and transfer of technology from north to south.

While Washington continues to debate, other nations of the world are seizing the opportunity to develop and launch their systems. We know that the Europeans have long had their regional system, likewise the Arabs will soon have theirs. And Indonesia early on developed a satellite system. All these systems provide services which INTELSAT could, and indeed does, provide in those regions. Yet, these systems have all been coordinated through INTELSAT without limitations or hesitations.

More recently, we have read in the trade press that Luxembourg, Sweden, Britain, France, and now Spain, are planning (and in the case of France, have launched) international satellite systems apart from INTELSAT. Obviously, world attitudes toward alternative systems are changing and new systems are springing up everywhere. The question is not whether there will be systems apart from INTELSAT. They already exist. The more appropriate question for those of us in this room, is whether our government should continue to stymie the use of modern communications technology by the countries of this hemisphere.

No one has opposed PanAmSat's Latin American domestic service -- not even Comsat. We have therefore requested the FCC to authorize this major portion of our system now. Last December, Congress amended the Communications Act to require the FCC to act within 12 months upon all applications for new technologies or new services, and it also established the presumption that such new services are in the public interest. Given this Congressional mandate and the lack of any opposition to PanAmSat's domestic service, we hope the FCC will soon move forward on our request. We believe the time has come for a prompt decision. We look forward to working with many of you here today to make PanAmSat's service a reality for the Spanish-speaking people of this hemisphere.

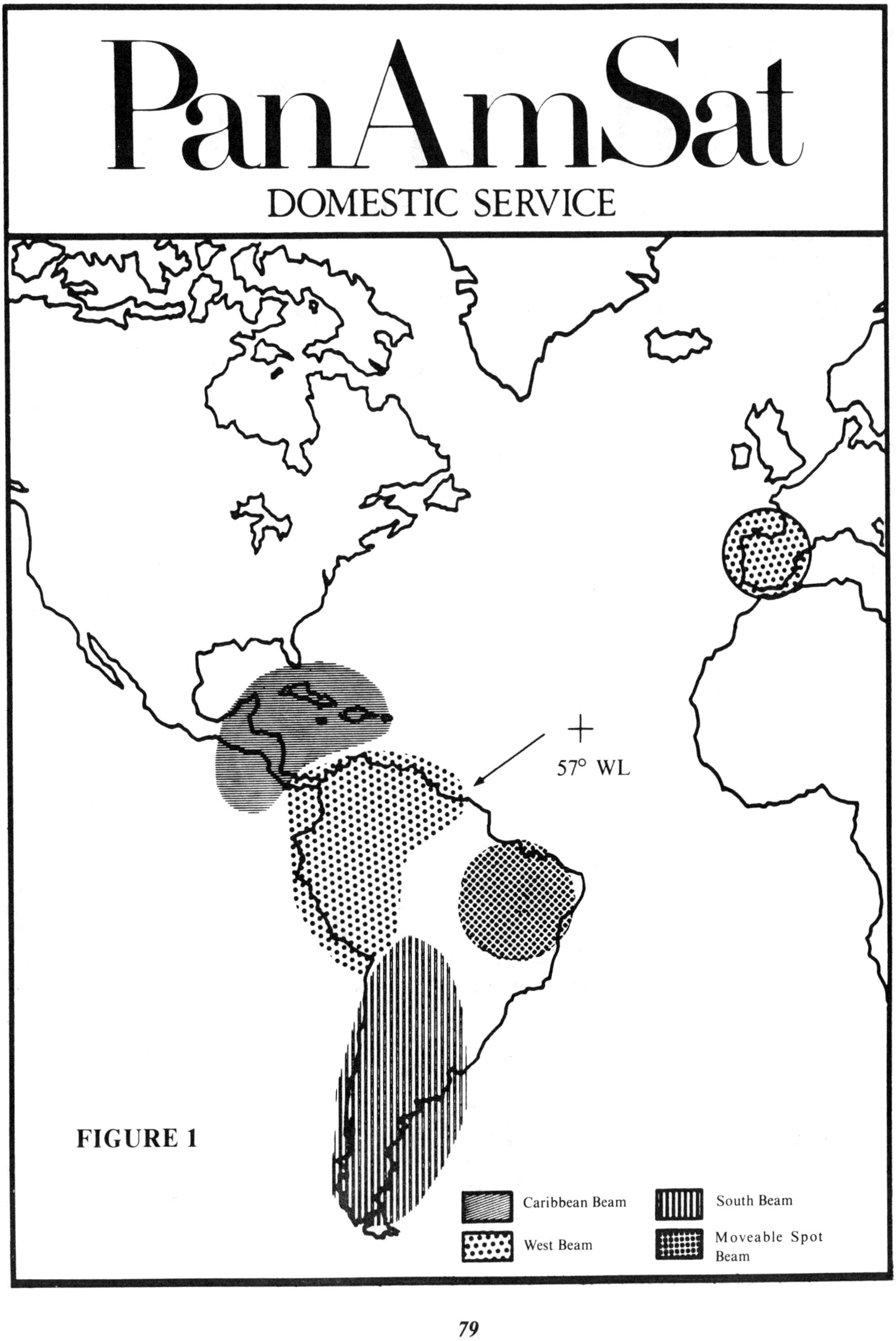

PanAmSat

DOMESTIC SERVICE

57° WL

FIGURE 1

Caribbean Beam

South Beam

West Beam

Moveable Spot Beam

PanAmSat
INTERNATIONAL SERVICE

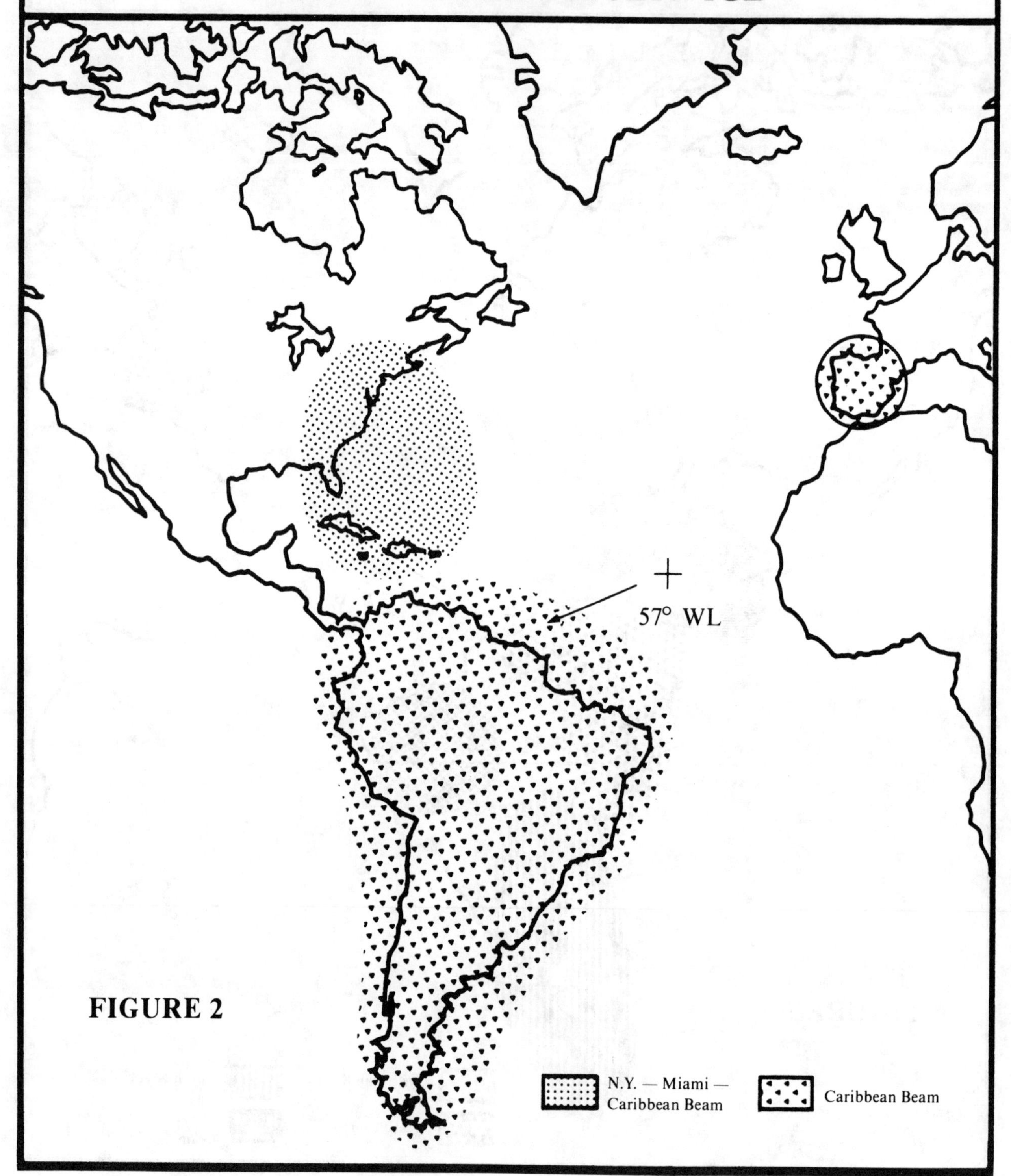

FIGURE 2

INNOVATIONS IN SATELLITE TECHNOLOGY APPROPRIATE FOR DEVELOPING COUNTRIES

by

Dr. Bruce Lusignan

Director, Satellite Communications Planning Center

Stanford University

What I'd like to do is to review for you briefly some particular technologies that are available widely. They have gone through the fire of the U.S. competitive market, and have survived very successfully. These technologies should form a part of the telecommunications capabilities of many developing countries. They've been proven in the U.S. industries. The prices that I am quoting today are quoted by U.S. manufacturers that have been established for a number of years.

First, I'd like to consider TV distribution systems. Currently you can buy a television receive only (TVRO) ground station in the U.S. for about $1,800. These are the type of TV receivers that you see atop motels and in people's back yards. To receive high quality television or rebroadcast quality television, you can go up to about $5,000. Within the U.S., there are projected to be about 200,000 of these types of receivers in the U.S. market this year, growing to about 500,000 next year.

gets seven to nine thousand circuits. It makes better use of the space segment. The systems of AT&T and RCA are being converted over to CSB trunkline capacity for major voice circuits between major cities. Again, that's a technology that, as far as a developing country is concerned, should be implemented rather than one of the older technologies that would make much less efficient use of the satellite.

There has been mention today of thin route telecommunication, and the great potential that is newly available. In the introduction of a satellite thin-route system that was implemented ten years ago in Alaska, 100 ground stations were installed with a total budget of $8 million. Even counting inflation, the current technology for these thin-route C-band systems is $40,000 to $75,000 per ground station. And those are off-the-shelf orders that you can place with two or three different U.S. companies at this time. Stations such as these were the basis of the thin route gateway stations for the rural areas of Mexico.

Another type of technology that is widely used in the U.S. but has not yet penetrated into the international market is point-to-point data distribution for applications such as weather, business data and so forth. Within the U.S. now these ground stations for receiving data use two-foot to four-foot diameter antennas selling for about $2,500. These data receive stations are well-established in the U.S.

These are C-band receivers for the domestic satellite systems. They are not the Ku-band receivers that have been touted for a number of years as having the cost advantage over the C-band. We've seen that when you combine the space cost and the ground cost, there is a significant advantage on the C-band systems, not the Ku-band systems as have been pushed for a number of years. The advantage of C-band was not apparent to many people in the U.S. until the recent financial difficulties of companies that put a lot of money into the Ku-band systems, and are now seeing that in fact they don't compete well with the C-band applications.

Another application that's important in telecommunications in any country is trunk telephony communications, communications of telephone systems between major centers. The technology in the U.S. that is now being implemented, the current modern technology, is companded side band modulation (CSB). The satellite telephone trunk systems are being converted from the older frequency division multiplexing (FDM-FM) systems in the U.S. over to CSB, the basic reason being that a typical transponder on a satellite can carry about 800 telephone circuits to 1600 telephone circuits with the older FDM-FM technology developed about 15 years ago. The newer digital technology, (time division multiple access) TDMA can carry something like one to two thousand circuits. The companded side band (CSB) technology instead of one to two thousand circuits

Two-way stations transmitting and receiving data will be introduced in the relatively near future with earth stations costing less than $7,000. The technologies that have been in place in the U.S. for two or three years are not expected in the international markets for another two or three years. The technology for the low data rate satellite distribution is spread spectrum technology, which has a significant cost advantage over other types of satellite technologies and is competitive with terrestrial means of distributing low data rate information.

Within the high data rate systems, hundreds of millions of dollars have been put into U.S. ventures to develop TDMA. At Stanford, we're finding for high data rate links, that the technologies that are most efficient use fairly standard single carrier, coherent modulation systems. These are transmitted through the satellites with separate carriers -- FDMA in other words. With this type of system, you can provide 2 to 7 megabits of data through the satellite with prices for the ground station with full redundancy of under $100,000.

All of these technologies have been developed and proven in the U.S. market and are available from successful U.S. companies. They are the types of technologies that ought to be under direct consideration by all countries in the international scene. Ideally, the prices on the U.S. market should be increased by 20% to 30% for the international market. However, a number of years of experience working for developing countries as well as the U.S. industries has shown me that there is serious blockage between the U.S. market, which has become extremely competitive and therefore very rapidly developing and cost competitive, and the international market. The competitive U.S. technology is not available on the international scene.

One of the biggest issues that I've seen in the ten years or so that I've been involved with developing countries is the problem of restricting access of competitive U.S. technology to the international market. The U.S. has recently created a great deal of publicity about the division of AT&T into many separate companies and allowing (in the regulatory sense) competition within the telecommunications industry. Actually, competition has been going on for about ten years, allowing connections of competitive equipment to telephone lines, and the "open skies" policy of allowing competitive satellites. Over this period, there has been an increase in the capacity and reduction in cost of technology by factors of two to four every five years in the U.S. But the improvement has not been available to developing countries. The difference in price for communications equipment between the U.S. domestic market and the international markets is on the order of two or three to one. This gap is by far one of the most serious problems in making telecommunications available to the developing countries, and using the technology best suited to those countries.

The technologies described earlier are established and competitive within the U.S. industry. They are well suited to most of the needs of the developing countries, but in fact they have not been made available. They are blocked by regulatory delays, planning delays and market protection. It's a problem that needs to be solved. A solution to this problem would go far toward making the most useful satellite technologies available to developing countries.

CANADIAN SPACE APPLICATIONS: MODELS FOR THE DEVELOPING WORLD

by

N.G. Davies

Director, Space Applications

Canadian Department of Communications

Most Canadians live in a narrow corridor of the country, along its southern border. Southern Canada is well developed, with many urban centers spread throughout the population corridor. Most schools, colleges and universities are located in this area. North of the corridor stretches approximately 80% of the country, with only 20% of the population. Population density in this region is small, and the further north, the smaller the centers, the more distant they are one from the other, and the more remote from urban Canada. In the high Arctic, there are some villages, but they are truly remote and isolated geographically. There are no roads between small settlements in the north; there are no telephone poles, no telegraph wires, nor microwave towers. This lack of terrestrial transportation and communication facilities is evident through much of the country in those areas where sparse population, great distances, mountains, permafrost, and barren-lands render such installations, if not physically impossible, certainly economically impractical. It follows that people who live in these areas are relatively underserved with respect to access to education, health care and other social services, compared to people who live in the urban areas.

Telecommunications has played an increasingly important role in linking all parts of Canada. Improving services to underserved citizens is a priority objective of the federal and other levels of government in Canada. A desire to improve the communications services led to the early involvement of Canada in the applications of satellite technology.

Canada's first satellite, Alouette I, a scientific research satellite, was launched in September 1962, to investigate the earth's ionosphere. Disturbances in the ionosphere can cause severe disruption to HF communications, then in widespread use in the northern regions of the country, and it was necessary to understand the ionosphere in order to cope with it and to overcome the problems. Alouette II was launched in November 1965, and two more scientific satellites, ISIS I and II, followed in January 1969 and March 1971. The Alouette satellites are no longer in operation, but both ISIS I and ISIS II continue to function. The exceptional performance of these satellites was encouraging, and the Alouette/ISIS scientific research program provided some answers to the question of what to do to provide reliable communications throughout all of Canada.

The focus of our space program swung from scientific research to communications satellites.

Telesat Canada, the domestic communications satellite agency, was incorporated in September 1969. The Telesat satellite ANIK A 1 was launched some three years later, in November 1972. ANIK A 1 had total Canada average, as did ANIK A 2, launched in April 1973, and ANIK A 3, launched in May 1975. With the ANIK A series, Canada became the first country in the world to have a geostationary domestic communications satellite system. This 6/4 GHz system provided for carriage of east-west heavy route telephone traffic and television program distribution through the country's seven time zones. Also, the system was used to extend network television and thin route telephone service to many communities in the north. A significant step had been taken to improve the delivery of services to underserved areas.

It was known, however, that southern television programs and telephone access to the southern telephone system would not be sufficient for northern residents. Access to other services, such as education and health care, was expected by them. Development of these kinds of services using conventional telecommunications delivery systems had begun in the south, and the results of individual projects were promising, but overall the progress in the country was slow. The required impetus for service development came with the launch, in January 1976, of the Communications Technology Satellite (CTS). This very powerful experimental communications satellite operated at 14/12 GHz. It was designed and built at the Department of Communications (DOC) Communications Research Centre under a joint Canada-United States program, and was named "Hermes" after launch. It was designed for a two-year mission lifetime, and the major portion of its life capacity was to be devoted to experiments in tele-education, telehealth, broadcasting and inter-community communications. Hermes lasted nearly four years, and the service development experiments were extremely successful. The Hermes experiments were short in duration, varying from a few weeks to a few months, and limited in operation to one to four hours a day, every other day. The experiments were essentially concept testers. Another step was needed, to test the viability of proposed systems.

The success of Hermes, with respect to both its 14/12 GHz technology and its application program, was evident early in the mission. Canada was thus able to plan the required follow-on program to maintain momentum. The Telesat ANIK B satellite, required for commercial use at 6/4 GHz, was specified to incorporate a 14/12 GHz payload for experimental use by the DOC. ANIK B was launched in December 1978, and was in service several months before

Hermes ceased operating, which assured continuity for the DOC program of service development. The ANIK B application program comprised a series of pilot projects, designed to provide educators and other participants with the opportunity to conduct pre-operational trials to determine viability of their proposed systems. Several Hermes experiments became ANIK B pilot projects, and a number of the pilot projects transferred to commercial satellite service.

To follow ANIK B, Telesat Canada placed orders for five more communications satellites. Three of these, the ANIK C series, are 14/12 GHz satellites; the other two are ANIK D satellites operating in the 6/4 GHz frequency bands. Telesat, therefore, could continue to upgrade its 6/4 GHz service, and implement a second Canadian satellite service at 14/12 GHz. Launches were scheduled to optimize satellite availability to match anticipated service demand. ANIK D 1 is scheduled for launch in November 1984, and ANIK C 1 in the Spring of 1985.

Consideration is now being given by Telesat Canada to the next generation of satellites, the ANIK E and ANIK F that will be needed in the 1990s.

NORTH-SOUTH ASPECTS

Canada is identified as being a developed country, and we usually think of the transfer of technological capability from north to south. However, in our case we live daily with the impact on our country of our large southern neighbor. We have to be conscious that with a population ten times larger than our 25 million people, activity within the United States will strongly influence our affairs.

In establishing Telesat Canada as the domestic communications satellite agency, provisions were made for the company to own the space and earth segments and to lease services only to the telecommunications carriers and the broadcasters. There were also provisions for the Minister of Communications to review major procurements by Telesat Canada to ensure the greatest possible participation by Canadian industry. The result was a minimization of the disruption of the established telecommunications carriers and significant development of the capabilities of the domestic industry. However there was very slow growth once the original requirements for trans-Canada telephone message service, distribution of the television and radio programs of the national broadcaster and thin route northern telecommunications services had been met.

The launch of domestic communications satellites in the U.S.A. by private companies in a competitive environment significantly changed this. The use of these satellites for the distribution of television programs by new entrants to the broadcasting scene, the super-stations, opened up new vistas for access to television programming with a number of impacts. While Telesat sought to lease earth stations to customers for television reception, the customers demonstrated that costs were lower for individual ownership; TVRO ownership had to be liberalized. While cable companies were licensed to distribute specifically identified Canadian and imported American television programs, the latter picked up from broadcast transmission and relayed through terrestrial microwave links, a wide variety of American television programs suddenly became available from U.S. satellites over the whole of the country. Ownership of TVROs for private entertainment purposes was liberalized.

Other events taking place in our southern neighboring country continue to influence strongly development in Canada.

EXAMPLES OF TRIALS AND THEIR RESULTS

The history of space experimentation in Canada led us to tackle the technical and institutional challenges of introducing new satellite telecommunications services in new frequency bands through experiments and trials in the areas of tele-education, tele-health, broadcasting, community communications, administrative services and technology development. Some examples are given drawn principally from the area of tele-education.

During the Hermes period, from January 1976 to 1979, thirty-seven experiments were conducted. Eight of these were experiments in tele-education, and six others included tele-education as an integral activity. During the ANIK B program, which began in December 1978 and concluded in March 1984, thirty-two pilot projects were conducted. Of this number, seven were tele-education projects and five incorporated a tele-education component. To date, three of the original Hermes experiments concerned with tele-education have made the transition via ANIK B to commercial satellite operations, and two more are possible; one additional ANIK B tele-education project is a possible candidate for commercial satellite operations.

The Ontario Educational Communications Authority, now better known as TV Ontario (TVO), is an agency of the government of the Province of Ontario. It was a firmly established and successful organization prior to the

FIGURE 1

ANIK-B PILOT PROJECTS (1980)

B: Broadcasting
C: Community communications
E: Tele-education
H: Tele-health
P: Administrative services

Hermes/ANIK B program, providing educational television programming to several urban centers in the province using terrestrial facilities. When the DOC made available the Hermes opportunity, TVO conducted three experiments:

1) July 1978 - a teaching/learning methods experiment involving four remote communities connected in an interactive mode (video receive, audio conference) to the service providing center (video transmit, audio conference).
2) August 1978 - an experiment to determine the learning needs of underserved areas - an interactive video teleconference between the service providing center and an underserved region of the province.
3) January - June 1979 - a direct broadcast by satellite (DBS) experiment wherein schools in four remote communities received educational television programs on TV receive only terminals via the satellite.

Results of the first two experiments were used in ongoing TVO planning and implementation of educational service delivery. Results of the third experiment in DBS were used by TVO as a baseline planning element for the development and implementation of an extensive ANIK B pilot project.

In the pilot project, forty-six underserved communities were equipped with TV receive only terminals. Low cost earth terminals (LCETs), a development of the DOC Research Centre, were installed at forty-two sites. The other four locations required antenna diameters larger than the 1.2 m and 1.8 m LCET size because they were outside the nominal satellite footprint. Receive terminals were installed at schools, homes, libraries, cable head-ends, master antenna TV systems, and one at a low power television repeater transmitter. Direct broadcasting by satellite began in September 1979. The project was a success from the start. When it concluded in September 1982, operations were continued by TVO using ANIK B on a commercial basis. In January 1983, operations were switched to the new ANIK C satellite. The TVO DBS network continues to grow through the addition of more receive sites in the province, and the inclusion of radio programming on video-sub carriers. One of the radio program sources will be the Wawatay Native Communications Society in northwestern Ontario, which conducted a radio distribution experiment using Hermes, and will transmit native language radio programs to 25 community radio stations in Indian villages.

TVO has just completed another ANIK B pilot project involving the delivery of Telidon teletext material via satellite to seven remote schools. This narrow-band (telephony circuit) trial provided for interactive data transmission so that residents of underserved areas could access large central data banks for computer aided learning purposes. It is anticipated that this tele-education system too will prove to be economically viable and move to commercial satellite facilities.

In the province of British Columbia, educational authorities conducted an eight-week Hermes experiment in the late fall of 1977 to explore the feasibility of satellite education for residents of communities that were isolated from urban institutions by distance and mountainous terrain. The remote communities were equipped with video receive, audio conference terminals, and the urban service providing center had a video transmit, audio conference terminal. One other remote site was allocated a TV receive only terminal. The experiment was successful, and was followed by an ANIK B pilot project.

The technical system for the pilot project varied from the Hermes experimental system in that more communities were equipped with TV receive only terminals. Audio interaction involving these sites was achieved through use of terrestrial telephone facilities. The project, which began operations in October 1979, was successful, and in 1980 the provincial government announced the formation of an educational communications authority, the Knowledge Network of the West, to continue and expand the satellite education activities of the pilot project. The DOC-owned interactive terminals were replaced by community-purchased LCET's, and other communities joined the network. In September 1982, when the pilot project concluded, the Knowledge Network transferred to commercial ANIK B operations, and then moved to ANIK C when it became available. There is a possibility that a second educational TV channel will be implemented by the Knowledge Network, perhaps in conjunction with other educational broadcasters in Canada.

In the late fall of 1978, an eleven-week Hermes experiment was conducted by Taqramiut Nipingat Incorporated (TNI), the communications society of the Inuit (Eskimos) of Arctic Quebec. The experiment connected nine villages in audio conference mode, through use of the 14/12 GHz Hermes satellite in tandem with the 6/4 GHz ANIK A system. Education in the community sense (skills sharing, cultural exchange, consensus building on issues on mutual concern, etc.) was one element of the experiment plan. Two Inuit pilot projects on ANIK B followed.

The TNI project provided four villages with video receive, audio conference terminals, and a fifth with a video transmit, audio conference unit. A television production studio was established at the fifth site. Programming was primarily in Inuktitut, the language of the Inuit, and again, general education was an incorporated component. The project ran from October 1980 through March 1981.

The second ANIK B pilot was conducted by the Inuit Tapirisat of Canada (ITC), the national Eskimo brotherhood, and operated during the period from August 1980 to May 1981. Education, both community and classroom, was integral to the trial. Two more Arctic television production studios were established for the ITC project.

The summary result of the two Inuit ANIK B projects, and the Hermes experiment, was the formation of another broadcasting network in Canada, the Inuit Broadcasting Corporation (IBC). The IBC transmits Inuit programming daily in the Arctic, at 6/4 GHz using transponders leased by the Canadian Broadcasting Corporation, (commercial ANIK C 14/12 GHz coverage is not available at that latitude).

Another native group, the Alberta Native Communications Society (ANCS), implemented a two-stage Hermes experiment from October 1976 to December 1977, and adult education was a primary program subject. Some classroom education trials were also included. Following this lead in the province, the Alberta Educational Communications Authority (ACCESS) conducted an educational ANIK B pilot project from March through May 1980. ANCS inputs were included in the ACCESS trials. Although the ACCESS project did not continue through to commercial satellite operations immediately, the experience contributed to a decision to augment the current terrestrial distribution facilities with satellite delivery commencing in 1985.

Memorial University of Newfoundland conducted a telehealth experiment on Hermes, March to June 1977, and part of the programming was designed for community health education, and professional and paraprofessional medical training. Since July 1980, the university has continued its work using ANIK B. Current activities include a telehealth satellite link to an off-shore oil drilling rig, where the work crews have expressed interest in tele-education opportunities. The possibility that this and other satellite circuits to the off-shore will become commercial in operation appears to be good.

RESULTS

The results of the Hermes/ANIK B program may be generalized as follows:

- Technology transfer of the 14/12 GHz satellite-based telecommunications systems from government laboratories to the user public was completed.

Awareness of the potential of satellite delivery systems, particularly at 14/12 GHz, was increased, and demands for commercial service offerings arose as the program proceeded.

- Technology development of 14/12 GHz earth stations occured in government laboratories and industry, in response to user needs for service delivery. New terminal designs, such as highly transportable television uplinks, low cost television receive only stations, and stabilized platform-mounted terminals, were among the technology development outcomes.
- Service development through satellite experiments and trials emerged as a viable process for market development, facilitation of technology transfer and stimulation of technology development.
- New services have been developed, and some existing services extended through satellite systems. (Other developments and extension of service have used terrestrial systems, i.e.: "Spin off" results of the Hermes/ANIK B program).
- The new technology and services developed and implemented during the Hermes/ANIK B program have impacted planning for satellite systems in other countries. For example, Canada has operated a form of low power DBS for four years; other countries will soon implement systems based on the Canadian model.

The specific results of the experiments and pilot projects are positive. Access on a continuing basis to greater educational opportunities has been extended to many communities, and the educational networks are expanding. Tele-education is a fact in Canada.

CONCLUDING REMARKS

In Canada, we have been fortunate to be able to carry out a program of trials using capacity on experimental satellites to prove out the new satellite technology and the potential for new services. The trials for new services were approached on an experimental basis that allowed organizations to gain a familiarity with the application and an understanding of the institutional implications before having to make a commitment to proceed with a service. The latter decision could be made on the base of much better knowledge of the benefits, the scope of the implementation and the costs.

It is likely that the involvement of a government agency is required, particularly in the early phases, as very few other organizations have the long term view or altruism to sponsor trials which may not have an immediate pay-off for the organization itself.

DISCUSSION

QUESTION:

What financial interests are backing PanAmSat and what services will it provide? Will it serve Cuba?

MR. RODRIGUEZ:

A group of investors have formed PanAmSat to file an application to the FCC. The financing will come from banks and from a public offering. The content to be transmitted by the system is the sole responsibility of the entity purchasing the transponder capacity. Two-thirds of capacity will be limited to domestic services in Central America, South America, and the Caribbean area. There may be PTT's that may want to extend their services, television stations that may want to use it for video transmissions, and common carriers that may use it for telex, facsimile, telephony, etc. International transmissions will be limited to video and audio distribution.

PanAmSat is aware of the need for this type of service, and there is a regulatory regime in this country which allows for flexibility which is not available elsewhere. Space capacity will be made available free of charge to universities in the hemisphere. PanAmSat expects this to be a condition of its license. PanAmSat does not have a marketing plan to sell transponder capacity to Cuba.

QUESTION:

What obstacles prevent technology transfer to developing countries?

DR. LUSIGNAN:

The primary obstacle is not protectionism in the Third World. The problem lies in the protection of the domestic industries in countries that market to those area. The international market in telecommunications is dominated by a number of very large companies which which are marketing primarily from Europe and Japan. The competitive ground stations I described in my presentation are not made by any of these international companies; they are made by competitive U.S. companies. The international market is a very restricted one in which mechanisms such as international planning bodies set the specifications for technical applications. This coincides with the highly regulated, less innovative characteristics of many European telecommunications industries. The international market is not competitive.

QUESTION:

Was there a provision for barter in the agreement between Brazil and Canada for purchase of Brazil's domestic satellite system?

MR. DAVIES:

A large component of the package was the satellites themselves plus the transfer of technology and the training that went with it.

PANEL 3: SATELLITES AND THE DEVELOPING WORLD

The paper prepared by Mr. Miguel Sanchez-Ruiz, director of Mexico's MORELOS satellite program, was presented by Mr. Raimundo Segovia, adviser for special projects to the Ministry of Communications and Transportation. Since 1970, Mr. Segovia has been on the faculty of the National Autonomous University of Mexico (UNAM) in Mexico City. In 1979, he joined the staff of the Director General of Telecommunications, and in the 1982, he was appointed as an advisor to the Minister of Communications for the MORELOS satellite program. The paper presented Mexico's plans for its domestic satellite, MORELOS, to be launched in 1985.

The second speaker was Mr. T.V. Srirangan, Member (Telecommunications Development) and Ex-Officio Additional Secretary to the Government of India, Posts and Telegraphs Board. Mr. Srirangan has had a long and distinguished career in telecommunications in India, and from 1976 to 1984 was responsible for all radio regulatory and spectrum management functions of India and for India's relations with the ITU. His career has included research and planning of satellite facilities, and he is currently responsible for the planning and development of the public telecommunications network in India, including the satellite (INSAT) based facilities. His paper was divided in two sections. He first reviewed the history of satellite communications in India, and described the domestic INSAT system and services. The second part, found in Panel 5 below, analyzed issues for the Space WARC (ORB 85 and 88).

The third speaker was Dr. Clifford Block, Associate Director for Educational Technology and Communications, the Office of Education, Bureau of Science and Technology, of the U.S. Agency for International Development (AID). During the past two decades, Dr. Block has initiated and supported through AID numerous innovative projects in the application of communication technology for development. Dr. Block's paper reviewed experience with the AID Rural Satellite Program which has assisted developing countries to apply existing satellite technology for developmental activities. Peru is using INTELSAT to reach isolated rural areas with two-way communications to support social service delivery and project management as well as public telephone service. Indonesia is linking remote university campuses via satellite for instruction and teleconferencing. An audio conferencing network operated by the University of the West Indies was also established under the AID program.

The final paper for this session was written by Mr. Walter Parker, President of Parker Associates in Anchorage, Alaska. Mr. Parker has made many important contributions to the development of communications and transportation in Alaska. Mr. Parker evaluated the education projects in Alaska on NASA's first Advanced Technology Satellite, ATS-1. While State Highways Commissioner, he played a key role in Alaska's efforts to bring reliable communications to its isolated villages. He has also been involved in the planning and implementation of the Learn/Alaska network, which brings educational programming to schools and homes throughout Alaska, and enables residents to participate in post-secondary courses through teleconferencing. His paper presented lessons for the developing world from the Alaskan experience.

KEY ISSUES IN SATELLITE COMMUNICATION: THE MEXICAN SATELLITE PROGRAM

by

Miguel E. Sanchez-Ruiz

Director General of Special Projects,

Mexican Ministry of Transportation and Communications

INTRODUCTION

Regulation of services that have a direct impact over the population must be based on careful planning criteria, which must satisfy the present and future needs of all countries, and, taking into account the particular needs and characteristics of each country, guarantee in practice for all countries, access to the geostationary orbit and frequency bands allocated for space services.

There are several planning approaches, each supported by different philosophies and arguments. For some countries, regulation will limit the potential growth of some systems that, eventually, will be needed in the future. For others, equitable regulation represents the possibility to assure minimum capacity for their future development.

On the one hand, developed countries are constructing societies based on information, that will demand high communication capacity (and computer power), especially in areas of long distance facilities (microwaves, satellite, fiber optics, cellular radio, broadcasting, etc.) For these demands, new approaches and more advanced technology are the answer. On the other hand, developing countries know the potential impact of communications on society and have plans to use them in order to support their development. However, they still have not satisfied demands for basic telecommunication services, such as telephony, telegraphy, radio, television, etc. And communication has a priority after food, medical assistance, education, housing, electricity, etc.

Each country has different communication requirements to accomplish specific programs in education, culture, entertainment, and information. Within each country, there is a variety of ethnic groups, with different languages, cultures, and social characteristics. Subsequently, implementation of the service planning must be based on rational requirements and with consideration of every country's reasonable needs.

KEY ISSUES

Based on the above I will address some of the main issues in the debate: "a priori" vs. "a posteriori" orbital planning approaches and "first come, first served" vs. guaranteed access options. How can we compromise or, what kind of concessions could we accept? The "a priori" approach might not be the best for the future, if we consider the possible technological changes, but we must consider also that technology could be applied for optimum use of the resources.

The "a posteriori" approach could mean a better use of the resources but only when they are needed; for some countries, this approach creates the uncertainty of having nothing in the future. It could mean accepting no guaranteed resources, in order to utilize them more efficiently, but eventually, benefitting only a third party.

These approaches combined with the "first come, first served" vs. guaranteed access options define another dimension to consider. The guaranteed access option is supported by developing countries because it represents the possibility of having something in the future. In a specific region, saturation of parking slots might occur for a band of frequencies, especially the less expensive. In such an event, countries that may need space capacity for the first time in that region, would face a problem, or at least a more expensive solution. On the other hand, there is the possibility of having resources assigned to a country without their use, while they might be needed for another country. That situation might appear totally unacceptable. Possibly the global policy will be between the extremes. Other compromise approaches may be a minimum guarantee of slots and frequency allotments.

The minimum guarantee approach may be to assign a basic capacity and then accept a posteriori planning for the non-assigned resources. Any new approach would reserve the basic capacity to countries that were not using any. In that way the access of all countries to the geostationary orbit should be guaranteed.

The slots and frequency allotment approach may consider a capacity for every country based on its requirements and its main characteristics (area, population, languages, topography, etc.). Every country could start using its assigned capacity according to an approved plan and standards; but future plans to improve the use of the resources will not be mandatory. This means that future technology might, eventually, be applied by each country when it is convenient. If a country required more capacity, it would have the possibility of reducing the satellite spacing, improving its antenna design and the earth stations, or some combination of these. Such a decision will cost more, but it could be introduced without concurrence of the rest of its neighbors.

Even the agreement on a global policy will require a great effort to arrive at a final recommendation. How will the different elements be weighted? What future services will be required? How will capacity be evaluated? What technology? What cost? Beyond these policies there are a lot of different positions and interpretations of the accepted goal of providing equitable access by all countries to the geostationary orbit. There are several interpretations of the statement. Apparently "equitable" has a lot of meanings, some of them too far apart from each other. Probably what we need is to learn the meaning of "equitable" and to apply such knowledge to telecommunications issues.

A conclusion of the above may be summarized as follows:

- international planning and regulations that establish the availability of frequencies or alloting frequency bands and orbital slots to each country;
- regulation and standards for new technologies, in order to provide a wide spectrum of solutions for terrestrial and spatial communication technology, that will offer a better perspective to all countries; and
- the need for more effective mechanisms for technology transfer.

THE MEXICAN SATELLITE PROGRAM

Based on the above basic principles and policies, the government of Mexico, through the Secretary of Communications and Transportation, decided to implement the MORELOS System that will also provide data transmission services, rural telephony services, educational TV services, etc. The satellites will be controlled through the (TT&C) control center located at Iztapalpa in the Mexico City area.

The technical characteristics of the satellites are summarized in Appendix A; for this discussion it is sufficient to say that the communication payload provides C band service with twelve narrowband 36 MHz transponders and six wideband 72 MHz transponders, as well as four 108 MHz transponders for Ku-band service (1,2).

A network of 200 communication earth stations is already operational in Mexico, using the F-1 INTELSAT IV satellite for domestic TV transmission and private telephone networks, that will be used with the MORELOS System.

Mexico has a population of more than 77 million inhabitants. Approximately 65% of them (50 million) live in 7,000 urban areas (localities with more than 2,500 inhabitants) most of them with TV, telephone and telegraphy services. Twenty million live in small villages with no telephony, telegraphic or TV services; this population is spread throughout the country in more than 100,000 villages of which there are 14,000 in the range of 500-2,500 inhabitants.

Based on the above, Mexico is planning to expand its telecommunication services through the use of the satellite for:

Rural Services:

- Rural telephony
- Rural clinics
- Oil exploration
- Electric power generation
- Educational television
- Agriculture
- Mining

Urban Services

- Trunk telephony
- National television networks
- Data transmission
- Private networks

The needs for these services and specific requirements are now being established through on-going studies to define the demand for telecommunication services in the country.

The impact of the telecommunications services of the MORELOS project can be seen by taking the present capacity of the federal microwave network for TV channels and observing that the capacity will be incremented by more than an order of magnitude. For trunk telephony also a substantial increment will be obtained. By these means the Mexican government is providing the country with a communications infrastructure that potentially will produce a variety of services that will help Mexico's program for a more equitable society.

REFERENCES

1. M.E. Sanchez-Ruiz et al. "The Mexican Satellite System." Proceedings of the 34th Congress of the International Astronautical Federation, October 10-15, 1983, Budapest, Hungary.
2. M.E. Sanchez-Ruiz et al. "El Sistema Morelos de Satelites: Una Respuesta al Reto". Proceedings de la Bienal CIME, 1984.

APPENDIX A: MORELOS SATELLITE SYSTEM SPECIFICATIONS

TABLE 1: COMMUNICATIONS SUBSYSTEM CHARACTERISTICS

Item	Characteristics	
	C Band	Ku Band
Channels	12 narrowband 6 wideband	4
Channel bandwidth, MHz	36 narrowband 72 wideband	108
Channel spacing, MHz	40 narrowband 80 wideband	124
TWTA output power, W	7.0 narrowband 10.5 wideband	19.4
Frequency bands, GHz		
Receive	5.925 to 6.425	14.0 to 14.5
Transmit	3.7 to 4.2	11.7 to 12.2
Antenna		
Receive	71 in. diameter reflector	Planar array
Transmit	71 in. diameter reflector	71 in. diameter reflector
Receiver	All MIC	All MIC
Channel gain control	0,3,6,9 dB commandable	0,3,6,9 dB commandable
Filters		
Input	Coaxial, pseudo-elliptical	Waveguide, pseudo-elliptical
Output	Dual mode, quasi-elliptical, waveguide	Dual mode, quasi-elliptical waveguide

TABLE 2: COMMUNICATIONS PERFORMANCE

Parameter		C Band	Ku Band
G/T			
Minimum receive antenna gain, dB		30.3	30.5
Antenna received noise temperature, K		270	270
Repeater noise temperature, K		590	610
G/T, dB/K		+1.0	+1.1
	Narrowband	Wideband	Ku Band
EIRP			
TWT output, dBW	8.5	10.2	12.9
Output loss, dB	1.7	1.2	1.1
Minimum transmit antenna gain, dB	29.2	30.2	32.2
EIRP, dBW	36.0	39.0	44.0

WHY ORBIT PLANNING: A VIEW FROM A THIRD WORLD COUNTRY

PART 1: INDIAN EXPERIENCE IN SATELLITE COMMUNICATIONS*

by

T.V. Srirangan

Member (Telecom. Development) P&T Board and

Ex-Officio Additional Secretary

to the Government of India

Ministry of Communications

INTRODUCTION

At the outset I wish to thank the University of Texas at Austin, for having given me the opportunity to participate in this International Symposium. I look forward to interaction with the very competent participants in this august assembly and to benefit from the same.

My presentation is organized in two parts. In the first, I would attempt to give a rapid, retrospective review of the Indian experience in Satellite Applications for Domestic Services. These have been covered in many other earlier papers and other publications, and hence I shall not go into details. In the second part, I wish to put forward some of my perceptions in regard to the issues that are slated for debate

---

* This paper presents the personal views of the author and they should not necessarily be construed as those of his Administration.

in the ITU World Administrative Radio Conference for the Planning of the Geostationary Orbit to be held in 1985 and 1988.

The success of EARLY BIRD in 1965 though for transcontinental communication, awakened interest also in the potential of geostationary satellites for domestic applications, particularly for countries with large geographical spread. Apart from industrially advanced countries like Canada, the USA and the USSR, developing countries like Brazil, India, China and Indonesia were considered as offering scope for domestic systems. The latter group of countries was in an early stage of development with very inadequate communication infrastructure for point-to-point telecommunication and broadcasting services which were recognized as vital needs for rapid socio-economic development. Studies were accordingly initiated in these countries, as also in several institutions in the USA, which was the leader in geostationary satellite technology.

In respect of India, the first such studies began in 1967. Prompted by the visionary zeal of the lake Vikram Sarabhai, the studies gathered quick momentum and were given a cléar focus. It was Sarabhai's conviction that India, with its vast size and population, and low level of economic development and literacy, should draw the benefits of satellite technology to make possible nation-wide television services to be introduced in one step for instructional, educational and related developmental purposes. There was at that time hardly any TV service in the country. It was Sarabhai's view that provision of TV services based on terrestrial TV transmitters linked by microwave links would involve considerable costs and time.

It took 15 years for Sarabhai's dream to be fulfilled in the shape of the INSAT-1 multi-purpose system. In the intervening period, many studies to determine the concepts and feasibility, techno-economic and cost benefits, identification of service needs, ear-marking of resources, etc., were gone through. Experimentation, system design and development, service demonstration, etc., were undertaken. Throughout this period, the objective remained sharp and clear: to reap optimum benefits of satellite technology for national development; and to achieve, progressively, substantial national self-reliance in this key and rapidly growing technology.

Attachments I-VIII summarize the important mile posts traversed by India since 1967. The course seems long and perhaps even meandering. But it has been worthwhile because of the many lessons learned and the opportunities created for achieving a substantial level of national competence in the different segments of telecommunications satellite technology, despite its unusual rapid growth. Given below are some of the very significant steps in which India has benefited by cooperative endeavours with several advanced countries, e.g. the USA, the USSR, France and the Federal Republic of Germany.

DIRECT T.V. BROADCAST SERVICES

SITE provided India a fine insight into the vital importance of programme software to carry an instructional/educational message to India's large and diverse population. The need to cater to local socio-cultural environment, language and ethnic variations, levels of agro-economic development, etc. and to ensure that programmes must have a distinctive local colours for effective developmental impact was realized. The need also to strike a good balance between entertainment and instruction and to combine them wherever feasible was also brought home. These have paved the way for subsequent actions in respect of TV broadcast services.

The maintenance organization to achieve operational reliability for the community direct reception TV sets in the remote and scattered locations with very little other infrastructure support posed many challenges. The indigenous development and manufacture of the hardware and the organization for their maintenance had to respond to the exacting requirements.

That the medium of television can be harnessed as a powerful instrument for socio-economic transformation of the country was proven beyond reasonable doubt and put at rest a continuing and controversial debate.

STUDIES AND THE STEP/APPLE AND OTHER PROGRAMMES

They confirmed that point-to-point telecommunication services are an essential ingredient in any domestic satellite system. They also established that such satellite based services are not a substitute for a well-designed and structured terrestrial trunk network, but function as an overlay facility offering improved overall reliability and flexibility in operations. The unique advantage of the satellite to extend telecommunications to remote, difficult to access areas including island territories was proved in practice. For short-term and urgent communication needs and emergency relief operations in any part of the country, the medium of the satellite was demonstrated to be most effective.

They showed that rapid provision of integrated TV-telecommunication facilities would become sinews for national integration and impart a sense of belonging to the inhabitants of islands and far removed parts of the country. Indeed, this should prove to be a great cementing force.

Studies and experience gained by India with meteorological satellite missions of other Administrations enabled the decision on inclusion of a meteorological package for synoptic observations of cloud cover over the country and its neighbouring regions to considerably enhance weather forecasting and storm warning capabilities. As India is a monsoon dependent economy, prediction and tracking of monsoons can lead to manifold benefits to agriculture, water and power resources management. Advance actions on the ground to minimize the devastating effects of cyclonic storms and floods, a common feature in India, would also become a practical proposition.

In total terms, opportunities were offered to get a better feel of many systems, technology alternatives, design, development and fabrication/manufacture of many items of hardware needed for the space and ground segments. A very substantial manpower development programme for both technical and managerial tasks was successfully embarked upon. An appropriate organizational structure for planning, implementation and operation of integrated nation-wide satellite based facilities has become possible.

THE INSAT SYSTEM

These should be considered as major benefits in the environment of a developing country. They have culminated in the planning and implementation of the INSAT I System with a multi-purpose mission, providing the following principal services:

1) Point-to-point telecommunication services of various types;
2) TV distribution; direct broadcast TV services for community reception;
3) Networking of sound broadcast transmitters;
4) Meteorological services based on day and night observations of India and its neighbourhood; weather related data collection from ground based platforms distributed all over the country and relayed to a central point by satellite; selective storm and flood warnings to target areas using S band.

In the one year period of operation of INSAT I-B, many studies have been undertaken to exploit the potential of INSAT 1. Several new services are also being planned. One of the most dramatic impacts has been the special programme nearing completion for expansion of TV coverage through a widely distributed networks of high power and low power terrestrial transmitters, linked via INSAT using both C and S band transponders. Some 150 transmitters, each with TV reception only facility from INSAT have been added in a space of 15 months with locally manufactured hardware. TV service on standard TV receivers for up to 70 per cent of the population has now become possible. In the rest of the country, especially the interior and sparsely populated parts, direct broadcast for community reception continues as the mainstay.

In sum, implementation of INSAT services is conclusive proof of the tremendous benefit potential of the medium for socio-economic and cultural development of the country. The importance of geostationary satellites to a developing economy cannot be questioned.

ORBIT/FREQUENCY COORDINATION FOR INSAT-1

An important element of the INSAT programme related to the international orbit-frequency coordination for obtaining suitable orbit locations. The effort beginnning in 1975 took over three years before registration with the IFRB could be done. And it was no easy effort to get the agreement of other concerned Administrations for the finally decided two orbit positions, namely 74°E and 94°E. This experience brought into bold relief the severe limitations of the extant radio regulations and procedures. It also revealed the risks and penalities which a "later entrant" to the orbit has to accept. In fact, the Government's decision in the INSAT System was very substantially influenced also by the desire to see that the opportunity for exploiting the benefits of a domestic satellite system was not lost because of the growing congestion in the orbit arc over the Indian Ocean, which is of interest to India. The very geography of this region renders this problem more difficult. That was why India decided to move the WARC-79 to decide on the planning of the orbit resource.

ATTACHMENT I

I. BACKGROUND

EARLY BIRD - 1965 - FOR INTERNATIONAL TELECOM SERVICES

IMMENSE POTENTIAL OF GEOSTATIONARY SATELLITES FOR DOMESTIC SERVICES AND NATIONAL DEVELOPMENT

INDIA AMONG CANDIDATE COUNTRIES

PRACTICALLY NO TV SERVICES IN INDIA: TELECOM NETWORK IN A RUDIMENTARY STAGE

VISION OF VIKRAM SARABHAI TO INTRODUCE IN ONE STEP NATIONWIDE TV SERVICES

ADVANTAGES IN TIME AND COST

ATTACHMENT II

II. CONCEPTUAL STUDIES

PROJECT ASCEND - STANFORD UNIVERSITY

PROJECT STRIDE - UNIVERSITY OF VIRGINIA

UNESCO STUDY MISSION

OTHER INDIAN STUDIES

FEASIBILITY STUDIES - 1968-70

PROJECT ACME - IIT KANPUR

INDIA - HUGHES CORPORATION JOINT STUDY

INDIA - G.E. CORPORATION JOINT STUDY

INDIA - M.I.T. (LINCOLN LABORATORIES) DESIGN STUDY

ATTACHMENT III

III. EXPERIMENTATION/SERVICE DEMONSTRATION

A. SITE

INDIA-NASA AGREEMENT IN 1969 TO SERVE AS A PILOT PROJECT FOR DIRECT TV BROADCAST FROM SATELLITE FOR RECEPTION BY COMMUNITY SETS

EXPERIMENT RESCHEDULED FOR 1975-76

KINDLED WORLDWIDE INTEREST

GROUND SEGMENT TOTALLY AN INDIGENOUS EFFORT

CONFIRMED BENEFIT POTENTIAL FOR INSTRUCTIONAL & EDUCATIONAL TV FOR NATIONAL DEVELOPMENT

EMPHASIZED CRITICAL IMPORTANCE OF RIGHT KIND OF PROGRAMMING SOFTWARE

ENABLED CONSIDERABLE COMPETENCE BUILDING FOR GROUND SEGMENT HARDWARE, SOFTWARE AND MANAGEMENT

ATTACHMENT III (cont'd)

B. SYMPHONIE TELECOM EXPERIMENTAL PROJECT (STEP)

AGREEMENT BETWEEN INDIA AND FRANCO-GERMAN SYMPHONY ADMINISTRATION

COVERED FOLLOWING:

* SCPC TECHNIQUES
* MULTIPLE SOUND CHANNELS FOR TV DISTRIBUTION
* INTEGRATION OF SATELLITE CIRCUITS WITH TERRESTRIAL SWITCHED NETWORK
* ROAD TRANSPORTABLE/AIRLIFT EARTH STATION TERMINALS FOR VARIOUS SHORT TERM/EMERGENCY SERVICES

SERVICE DEMONSTRATIONS; COMPETENCE BUILDING

EXPERIMENTS CONDUCTED FOR TWO YEARS 1977-79

ATTACHMENT IV

IV. PROJECT APPLE - 1981-83

(ARIANE PASSENGER PAYLOAD EXPERIMENT)

3 AXIS STABLIZED SPACECRAFT DESIGNED AND FABRICATED IN INDIA WITH ONE TELECOM C-BAND TRANSPONDER

LAUNCHED AS CO-PASSENGER ON ARIANE DEVELOPMENT FLIGHT IN JUNE 1981

LOCATED AT 102°E ORBIT POSITION

USED FOR PROVING SPACECRAFT'S DESIGN AND HARDWARE

USED ALSO FOR FURTHER TELECOM EXPERIMENTS INCLUDING TDMA, DATA LINKS, VIDEO CONFERENCING, ETC.

SHUT OFF AFTER 2 YEARS OF SERVICE

ATTACHMENT V

V. IDENTIFICATION OF SERVICE NEEDS/PRIORITIES/COSTS

INITIALLY PERCEIVED MAINLY FOR TELEVISION WITH TELECOM AS PIGGY BACK TO ENABLE REVENUE YIELDS

TERRESTRIAL OR SATELLITE ?

NATIONAL CONSENSUS ON INTEGRATED TERRESTRIAL-CUM SATELLITE BASED TELECOM SERVICES FOR MAXIMUM BENEFIT AND FLEXIBILITY - 1972

PLANNING COMMISSION TASK FORCE - 1973
IDENTIFIES PRIORITIES AS: REMOTE SENSING/METEOROLOGY; TELECOMMUNICATIONS; TELEVISION

COST ESTIMATED AT $400 MILLION WITH COMMUNITY SETS IN 250,000 VILLAGES

DECISION DEFERRED DUE TO PAUCITY OF FINANCIAL RESOURCES

ATTACHMENT VI

VI. INSAT SYSTEM DECISIONS

SUCCESS OF SITE RENEWED INTEREST FOR GOVERNMENT APPROVAL FOR AN OPERATIONAL SYSTEM

TREND TOWARDS ORBIT CONGESTION IN INDIAN OCEAN REGION CAUSED CONCERN

HIGH LEVEL INTER-AGENCY COMMITTEE STUDIED IN DEPTH SERVICE NEEDS, COST, TRADE OFFS, PRIORITIES AND PROVEN TECHNOLOGIES - 1976

RECOMMENDED A MULTIPURPOSE CONFIGURATION FOR TELECOM, S-BAND TV AND METEOROLOGICAL SERVICES

APPROVAL OF GOVERNMENT - JULY, 1977

ORBIT POSITIONS 74° and 94° COORDINATED AFTER EXTENSIVE AND DIFFICULT BILATERAL DIALOGUES

ATTACHMENT VII

VII. INSAT LAUNCH & OPERATIONS

INSAT 1A LAUNCHED - APRIL 1982

- FAILED IN ORBIT SEPTEMBER 1982

INSAT 1B LAUNCHED AUGUST 1983

- OPERATIONAL SINCE OCTOBER 1983

INSAT 1C DUE FOR LAUNCH MID-1986

ATTACHMENT VIII

VIII. INSAT STATUS AND BENEFITS

CONSIDERABLE IMPETUS TO GROWTH OF ALL THREE SERVICES

OFFERING OVERLAY TELECOM CAPACITY IN ARTERIAL TRUNK ROUTES

REMOTE AREA RELIABLE COMMUNICATIONS

- CONSIDERABLE SPURT IN DEMAND

QUANTUM LEAP IN NATIONAL TV COVERAGE

- WITH ALL ITS SPIN OFFS

ENHANCED CAPABILITY FOR WEATHER FORECASTS, STORM AND FLOOD WARNING, ETC.

SUBSTANTIAL PROGRESS IN ADDITIONAL HARDWARE AND SOFTWARE CAPABILITIES FOR ALL SERVICES

- TOWARDS SELF-RELIANCE

SECOND GENERATION INSAT FORESEEN FOR END OF DECADE: SYSTEM DEFINITION NEAR FINALIZATION; SPACECRAFT TO BE INDIGENOUSLY BUILT

SATELLITE LINKAGES AND RURAL DEVELOPMENT*

by

Dr. Clifford H. Block

Associate Director of Educational Technology

and Communications, S&T/ED

Agency for International Development

Washington, D.C.

INTRODUCTION

This paper describes key elements of our experience to date in a venture called the A.I.D. Rural Satellite Program. This program is a cooperative effort among the United States, a number of Caribbean countries, Peru, and Indonesia, to explore the potential of satellite communications for rural and institutional development. University of Texas people, I must note, have been central to this venture. Professor Emile McAnany worked in the earliest planning, and Professor Heather Hudson was both a valuable consultant and an indefatigable prod to the United States Government to undertake the program. Later, after we did decide to undertake it, we turned to Texas again and asked former Professor Robert Schenkkan to come to Washington to implement these Texas-sized ideas as manager of the project.

---

* These remarks are those of the author, and do not necessarily reflect those of the United States Agency for International Development

The A.I.D. Rural Satellite Project (see Figure 1) has a number of components but the key is a series of pilot projects, each centered on the utilization of telecommunications delivered by satellite: first, utilization by institutions crucial to rural education, agriculture, and health; second, utilization by individuals through access to rural telephones.

This is a progress report of an activity whose pilot stage is two-thirds complete; we plan to continue A.I.D. support for the pilot efforts through September, 1986. I intend to share what we have learned and what we haven't learned, and to be very candid about those problems that have been greater than expected - political, technical, and not the least, budgetary.

Why may this experience be important at this time? First, reliable communications may be just that catalyst needed to spur lagging rural development efforts, through making possible dramatic improvements in rural administration, information, and training, and satellite technologies remain the most practical vehicle for providing such communications within the predominantly rural developing world. Second, the long awaited day of the higher powered domestic and regional satellite is now here, for much of the world. When one adds up the populations

FIGURE 1

AID RURAL SATELLITE PROGRAM COMPONENTS

* PILOT PROJECTS
  - TECHNICAL ASSISTANCE FOR PLANNING, IMPLEMENTATION, OPERATION, SYSTEM ENGINEERING
  - HARDWARE
  - TRAINING
  - EVALUATION
* POLICY STUDIES
  - FINANCING CRITERIA AND MECHANISMS FOR RURAL TELECOMMUNICATION SYSTEMS
  - SATELLITE SYSTEM SURVEY
  - OTHERS
* INFORMATION SERVICES
  - NEWSLETTER
  - LIBRARY/REFERENCE SERVICES
  - A/V SUMMARIES OF PROJECTS
  - SEMINARS/CONFERENCES
* TRAINING
  - SURVEY OF TRAINING OPPORTUNITIES
  - TRAINING NEEDS ASSESSMENT FOR RURAL SATELLITE PROJECTS
* EVALUATION
* HARDWARE MANAGEMENT
  - STATE OF THE ART MONITORING
  - SITE SURVEY, SYSTEM DESIGN, PROCUREMENT SUPERVISION
  - INSTALLATION, CHECK-OUT, OPERATION SUPPORT
* PHOTOVOLTAIC ENERGY (NASA RESEARCH PROGRAM)
* SMALL EARTH STATION R & D (ITS RESEARCH PROGRAM)

that will be covered by INSAT, ARABSAT, the Mexican satellite and the Brazilian system, all to be operational by 1986, one finds that over a billion people will be a major step closer to basic communications. Perhaps lessons from this project can be of help in utilizing these new capabilities, as earth station coverage grows in these regions.

Finally, our experience may help countries determine to what extent they might usefully invest in the terrestrial infrastructure needed to utilize satellite communications in their rural areas. Rural utilization is not automatic. When satellites first go into operation, countries tend quite naturally to focus on their use for high-density inter-urban telephone linkages, and for television broadcasting, rather than for rural telephony. Although such priorities are not irrational, we hope this project will show such great developmental usefulness for the rural areas that some nations will find a basis for accelerating rural coverage through earth station networks. If combined with assistance to rural institutions in making full use of this new capability, an important new development tool may be at hand.

What has been done in the A.I.D. Rural Satellite Project? Principally, work has taken place in three pilot sites. We have provided <u>technical assistance</u> and <u>training</u>

to support those pilot projects, principally through the Academy for Educational Development. We are evaluating the pilot experiences, through work done by Abt Associates and Florida State University. We carry out an information activity which includes a newletter called "Uplink", a periodical on our progress and travails that goes out to about 4,000 people around the world. We've carried out a bit of technological R&D on small earth station designs, through the Institute for Telecommunications Sciences, and we are about to test a solar-powered small earth station in Indonesia, developed by NASA and Hughes Aircraft. But, primarily, we have focussed on trying to determine how to help user institutions in the pilot sites to utilize this new communications capability.

Technologically, we have focussed, unlike most other previous projects, on audio applications and on applications of an interactive nature, specifically audio conferencing and community telephony. The activity in Peru includes both of these applications, while the other projects focus on audio conferencing links among distant institutions.

THE PERU "COMMUNICATIONS SERVICES" PROJECT

In Peru, a pilot project began operation in November, 1983. The project is located in the Department of San Martin, on the eastern side of the Andes, a rapidly growing region of new settlement, with very limited infrastructure.

It is now a seven-community network, comprised of three earth stations, plus inter connections from one earth station to four nearby villages by VHF. The system provides both a conferencing network and community telephones in each village. (See Figure 2)

The intention is to integrate satellite telecommunications as intimately as possible into the processes of rural development and thus multiply the availability of local expertise, improve administration, and establish reliable linkages among business, farmers, markets, and suppliers. The earth stations are Harris Corporation earth stations designed for the project, using 6.2 meter antennas, costing about $200,000 each, and operating off an INTELSAT Global Beam. Each is currently capable of handling 4 circuits, and is easily expandable to go up to 8 circuits. Since the Global Beam is the least powerful INTELSAT service, these earth stations had to be relatively large, although far smaller and less costly than the typical INTELSAT "B" station otherwise generally in use, which costs about $1 million. Higher-powered services will make future stations less costly.

The project is being carried out in Peru by ENTEL Peru, that nation's telecommunications authority. An important institutional development resulting from this project is that ENTEL/Peru has established groups within its

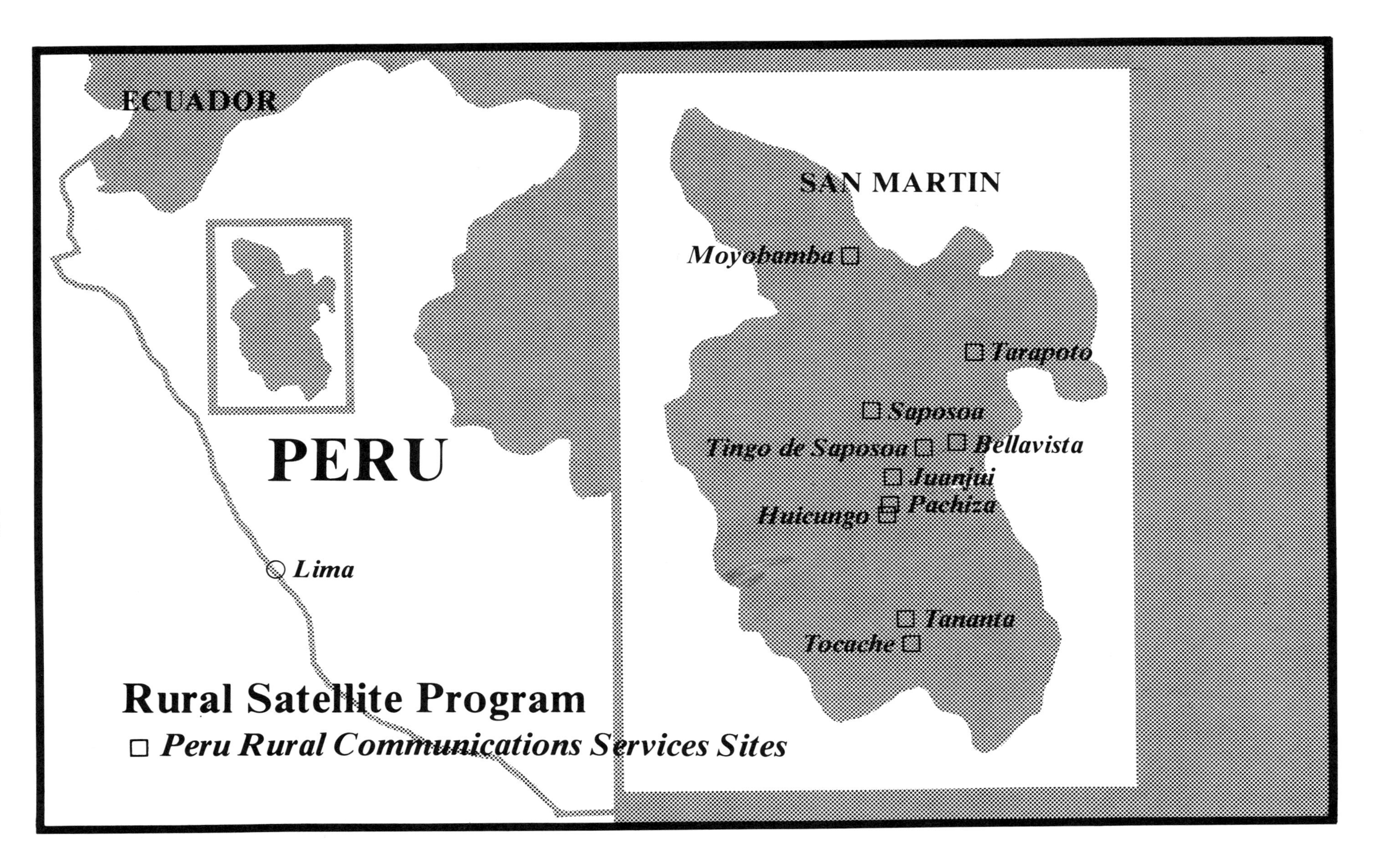
ECUADOR
PERU
Lima
SAN MARTIN
Moyobamba
Tarapoto
Saposoa
Tingo de Saposoa
Bellavista
Juanjui
Pachiza
Huicungo
Tananta
Tocache
Rural Satellite Program
□ Peru Rural Communications Services Sites

FIGURE 2

organization dealing not only with the technology, but also with applications planning and evaluation, groups which work with local communities on how to utilize this capacity. This is a most important sign for sustained development of rural telecommunications in Peru.

In order to promote innovative uses, ENTEL is offering each of the participating local agencies a specified number of toll calls daily that are free of charge; in addition, each is provided free use of the audio-conferencing circuit. In addition, there are in each community public phone offices which charge commercial tariffs. Our evaluation is analyzing not only the impact on particular sectors such as agricultural and health services, but in addition is looking at any impact on overall economic development. Can a nation, in fact, spur economic development, as we have all been asserting for a very long time, simply by providing effective telecommunications?

What is the status of the system now - a year after initiation? The most positive result of all is that user acceptance has been tremendous, with long lines at the community telephones. (The down side of that is that the lines are sometimes long because the system doesn't always work!) The audio-conferencing system is heavily and increasingly used; there are, thus far, an average of twenty to thirty audio telephone conferences per month, and the number is growing. The system's use in health is rather like the famous Alaskan model, in that indeed the only trained physicians in the area, in two regional hospitals, regularly network with community health providers, assisting them with diagnosis and treatment, often with patients actually present in the villages. Educational and agricultural applications are also coming along, providing training and administrative support for extension workers and teachers. In sum, a wide variety of community uses are underway, with enthusiastic acceptance.

The equipment in Peru, as with the other projects, has proved somewhat troublesome. It is a somewhat noisy system, and it hasn't always functioned reliably. We need to keep in mind that this sort of system, designed to operate in a rural area and using conferencing techniques, does present different technological demands that will continue to require attention. Power-line variations, problems with the innovative earth stations and with their interface with the terrestrial system and line-balancing problems for the conferencing function have all had to be worked at. Even with such problems, however, use has steadily grown, illustrating the high perceived value of a very basic system to areas otherwise unserved.

The Peru Rural Communications Project is currently the only one in the developing world that is testing telecommunications as a multipurpose rural development tool. Assumptions that reliable satellite telecommunications can be a catalyst for rural development--to reduce the administrative difficulties of providing rural services, to provide economic information links for farmers and entrepreneurs, to attract new investment, to develop regional cohesion--are being put to an important test in this project. The management of this "social experiment" by Peru's telecommunications authority itself helps to ensure that the project will address the questions essential to future decision-making and investment choices, and importantly, helps to ensure that the project's development orientation will continue on an operational basis after the initial project period. As evaluation data becomes available starting in 1985, we expect to learn a great deal.

THE UNIVERSITY OF WEST INDIES DISTANCE TEACHING EXPERIMENT

In the Caribbean, quite a different model is in operation. This is a program being carried out by the University of the West Indies linking its three main campuses in Jamaica, Barbados, and Trinidad with smaller, "extra-mural" campuses on the island nations of St. Lucia, Dominica, and Antigua (see Figure 3). This project has been operating since March, 1983; it has evolved as a leased telephone-linked system, with the satellite link being simply a standard INTELSAT link between Jamaica and Barbados. (Initial experimentation was with very small

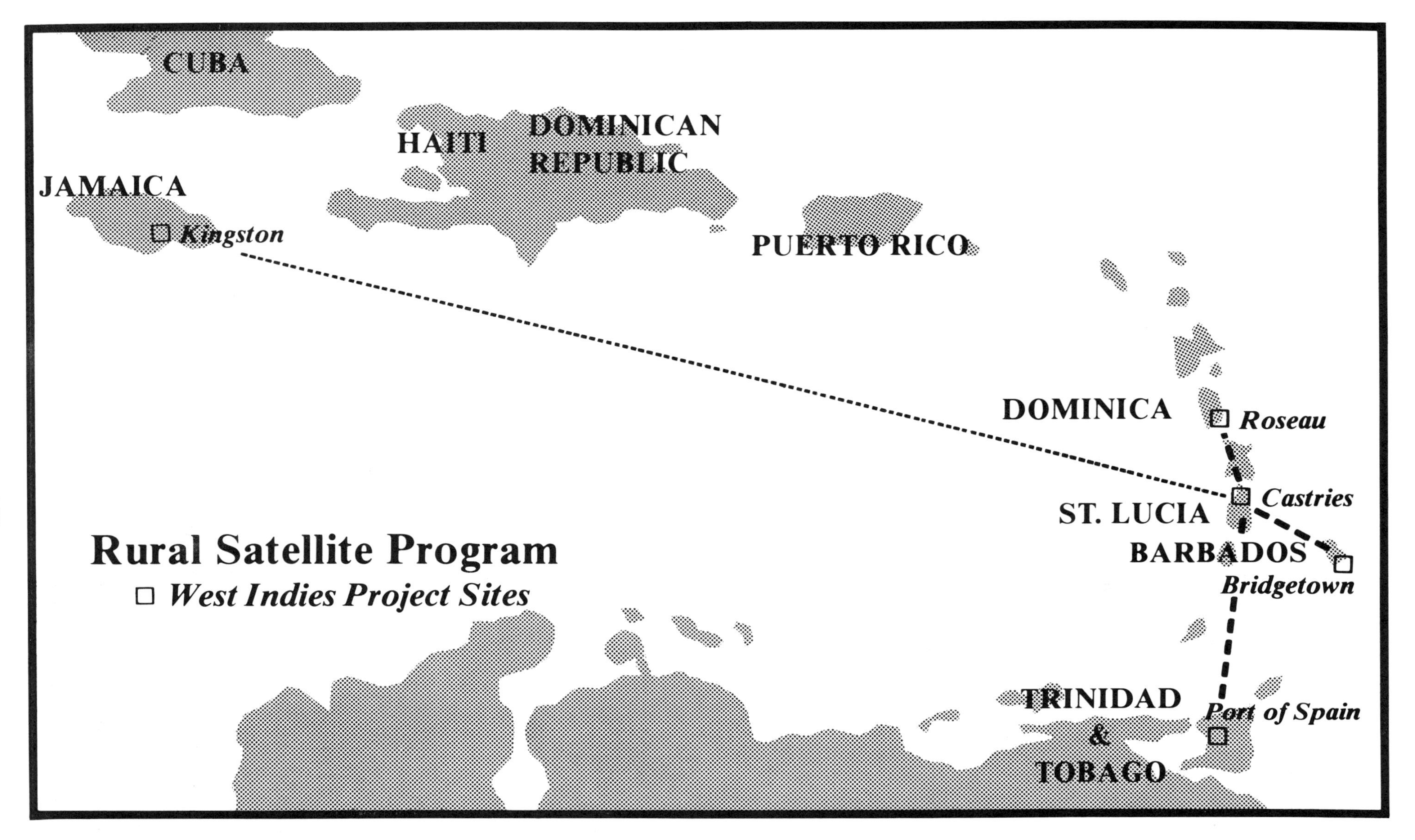
CUBA
HAITI
DOMINICAN
REPUBLIC
JAMAICA
Kingston
PUERTO RICO
DOMINICA
Roseau
Castries
ST. LUCIA
BARBADOS
Bridgetown
TRINIDAD
&
TOBAGO
Port of Spain
Rural Satellite Program
West Indies Project Sites

FIGURE 3

earth stations using NASA's ATS-6 and ATS-3.) It could, however, develop in the future into a system using small earth stations.

This project exemplifies a distance teaching concept; that is, to share limited intellectual resources more widely and to provide university opportunities for students who live on the small islands, who otherwise would have to go to one of the main campuses, with all the expense involved. The University of the West Indies has for some time had procedures where students could take first-year exams while not in residence, but the difficulty was that education was almost entirely self-instructional, and failure rates were extremely high. Now, there is both formal coursework and tutoring available over the interactive audio system. Currently, basic first-year academic courses are being offered in economics, sociology, accounting, history of the Caribbean, and mathematics and statistics. In addition, a series of in-service course is being offered for teachers, in the fields of teaching for the hearing impaired, mathematics, and science. Fifteen months of successful participation earns a university certificate in education, as compared with 12 months in residence, in a more costly and non-earning capacity. The results are encouraging -- at the end of the first few months the first year examination in one subject showed a doubling of the pass rate. So, the system seems to be having an effect on quality and also is increasing access substantially.

Another most encouraging development is that there are many special uses occuring, particularly in areas where there aren't really enough people to design a formal program and bring people for training. Recently, for example, UWIDITE collaborated with the Caribbean Food and Nutrition Institute to offer an eight-hour per week, two-month long course for community workers in agriculture, education, health, and community development. The course was created to update food and nutrition concepts, create awareness of nutrition problems and measures for overcoming them, and strengthen skills in working effectively with the community to promote good nutrition. A second outreach program offered a 20-hour course on training for trainers of day-care personnel. The region's family planning experts have used the system extensively to train community health workers; Caribbean-wide laboratory technician training is going on, and now training for agricultural workers is slated to begin. In a series of consultations that began late in 1984, the Cardiology Group of the University Hospital of the West Indies has taken advantage of the network's slow-scan television capability to transmit electrocardiograms and x-ray photographs.

Demand for the use of the UWIDITE network has increased rapidly over the past year, and scheduling has become a problem, particularly with the addition of more programs catering to working adults. As distance education gains

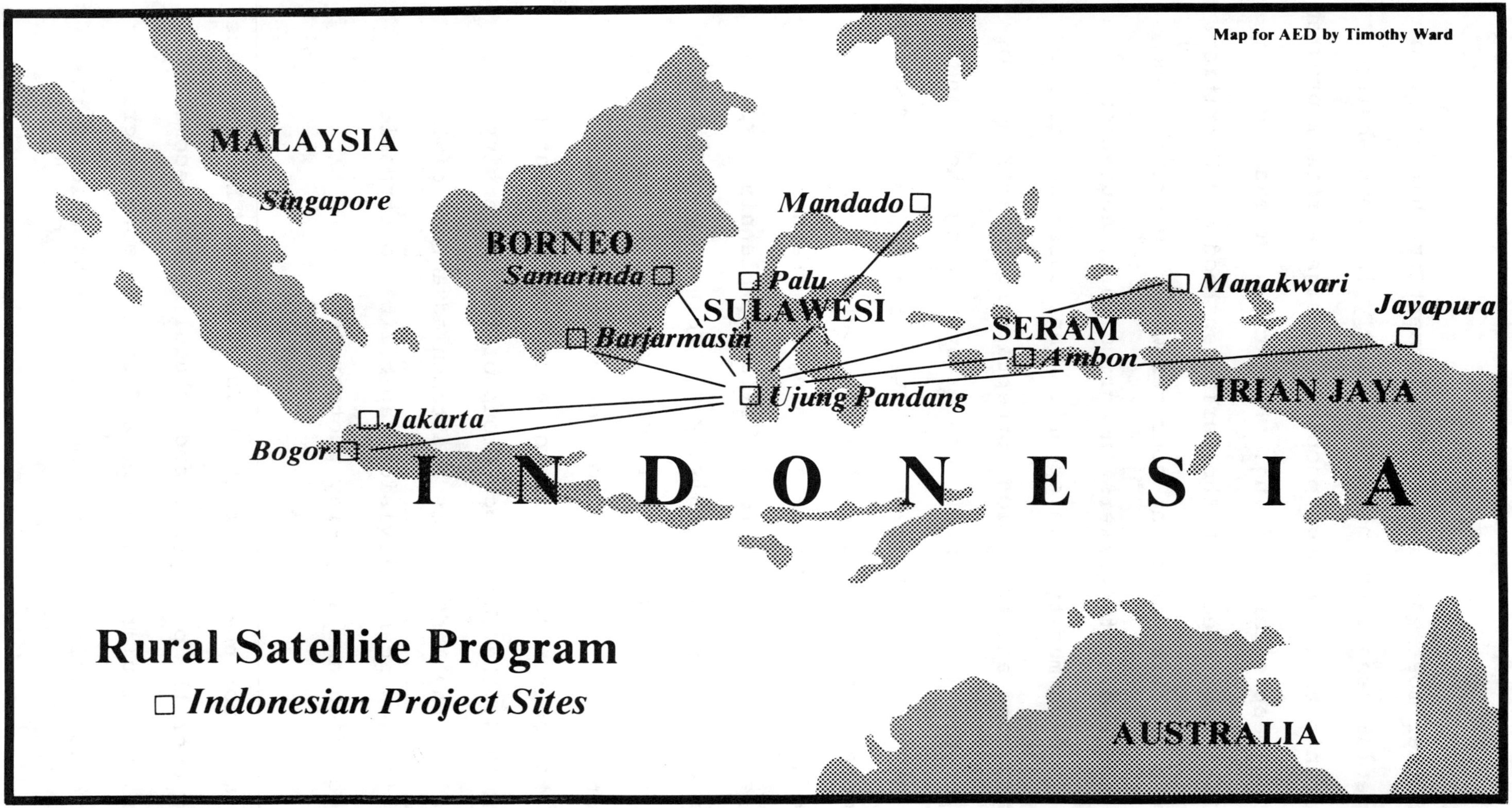

Map for AED by Timothy Ward

MALAYSIA

Singapore

BORNEO

Samarinda

Barjarmasin

SULAWESI

Palu

Mandado

Ujung Pandang

SERAM

Ambon

Manakwari

Jayapura

IRIAN JAYA

Jakarta

Bogor

INDONESIA

AUSTRALIA

Rural Satellite Program

□ *Indonesian Project Sites*

FIGURE 4

recognition in the Caribbean, more countries are asking to be included in the network, and the university is looking for ways to expand the network to reach a larger audience. UWIDITE has already had an enormous impact in the Caribbean and has won acceptance and support both inside and outside of the University as a valuable experiment.

INDONESIA: THE EASTERN ISLANDS UNIVERSITY ASSOCIATION

In Indonesia, since October, 1984, there has been in operation a somewhat similar system, covering a huge expanse of ocean using Indonesia's PALAPA satellite, which has been in place since 1976 and has until now been used largely for conventional commercial telephony, telex, and broadcasting, with few special social service offerings.

During the PALAPA system's planning in the early 1970s, there was strong government interest in using the satellite system to support education. While the expansion of Indonesia's basic telephone system took first priority, PERUMTEL and a number of development-oriented ministries maintained that interest, encouraged by A.I.D. and others.

The PALAPA satellite offers unusual opportunities for extending development communications. It is a high-powered system: small, inexpensive earth stations can be used, offering the possibility of widespread development of communication services in the most remote areas of the archipelago.

An agreement was ultimately worked out to provide A.I.D. assistance, under the Rural Satellite Project, as a means of strengthening a group of new universities and teacher training colleges located on the islands of Kalimantan, Irian Jaya, Seram and Sulawesi. (See Figure 4) Satellite project classrooms in Indonesia, as in the Caribbean, are outfitted in an audio-conferencing mode. There, we are in addition experimenting with an audio-graphic system, one developed by the British Open University, (known as the "Aregon" system in this country, "Cyclops" in England). It uses a telephone bandwidth circuit to permit a lecturer to write on a graphics tablet and have his lecture notes displayed on a television screen. Aregon is fully interactive, so that students or tutors at local sites, can, for example, point to parts of a diagram or respond to questions "on the blackboard." One of the reasons we went to this kind of system is that teachers were telling us in all of the pilot project sites that audio would not be enough, that they needed some kind of visual presentation. I frankly don't know if that is really going to be true after teachers have had greater experience with the system, because I know they are finding, in the West Indies, that audio, particularly in the interactive mode, is a very flexible and easy instrumentality to use. Technically, it is easier to handle and more reliable than the relatively new audiographics technology. But this experimental period with narrow-band visuals should answer a number of questions of importance to distance teaching.

The capability of the satellite also provides a link between the faculties of the new universities scattered throughout eastern Indonesia with experts at the Bogor Institute, a world-renowned regional research institute in agriculture located on Java; thus, faculty upgrading is made possible. It should also be noted that the system in the Eastern Islands represents a potentially new model for institutional development, because each of these new universities now has an opportunity to specialize in one particular area -- rice culture, soil science, or whatever -- rather than to try to develop immediately the full range of faculty capabilities, in a situation where Indonesia is short of thousands of trained faculty. It may illustrate a potential for profound impact on the way we go about developing institutions, when we take full advantage of the new communications capabilities.

Experience with the Indonesia system is still relatively slight, since the first usage began only in October, 1984. However, already:

* Students formerly receiving statistics instruction from instructors with minimal training are now studying

statistics with one of Indonesia's best-trained statisticians.

* The rectors of the nine universities making up the Eastern Islands Universities Association now meet together every month instead of three times a year as before--without even leaving their campuses.

* Managers of a special graduate agricultural program conducted by two universities located 1,000 miles apart are now meeting once a week to explore ways in which they can enhance their program and also reduce the time lecturers and students spend away from their home campuses.

On the technical side, local line noise initially hampered the system, but has been significantly overcome. The satellite itself serves as a teleconferencing bridge. While the use of the satellite as a bridge is more cost-effective than a terrestrial bridge, it compounds any noise problems by broadcasting the noise of one site automatically to all other sites. A transmit cut-off system is being installed so that noise transfer can be limited.

Another major question is the durablity and reliability of the electronic equipment in Indonesia's heat and humidity. A number of equipment failures have occurred, despite attempts to "weather-proof" the equipment. Experience to date underlines the ongoing concern with repair and maintenance procedures and management.

Early in 1985, experimentation will begin with a solar-powered, 5 meter earth station in the remote area of Wawatobe. This station has been designed to operate on only 500 watts of total power. The implications for rural communications of small earth stations using solar power are apparent.

During the next two years, evaluation will focus on a number of key issues: the appropriatenesss of the current equipment mix (audio, facsimile, and telewriter); the value of student interaction in this society; methods of course development appropriate to this context; and many other questions of affordability, cultural acceptability, and utility. Systematic data is being collected that should prove of considerable interest.

PROGRESS AND PROBLEMS

As you can tell, I am most enthusiastic about our progress and our potential. I am going to be quite candid however, about some of the distressing things we have learned. One of them is that this kind of application is not by any means always politically or economically acceptable. We designed three such pilot projects that never got started. The first was blocked by a perception in key quarters that this technology was inappropriate for the level of development of certain poor societies. That may in fact have been true, but we were unable to put the assumption to a test, and that is unfortunate. The second was eliminated when foreign assistance to a huge river basin development project in Africa was reduced, and communications was simply deemed of too low a priority compared to irrigation, roads, and the like. The third dropped by the wayside when a government communications ministry concluded that if the project succeeded, a continuing subsidy would likely be required to maintain the telephone system for a rural province. Although promises were made by user ministries - agriculture, rural development, education - that they would make up the budgetary difference, those were not compelling promises. The communications ministry in that country concluded that they would be running at a perpetual deficit there and, I think, they were really afraid of success. Had this succeeded, the political demand for this kind of rural communication service around the country would, they feared, have overwhelmed them. So, there has been some very substantial, and not unfounded, resistance. And then, there are other countries, as you have seen, that have taken all this in stride, have worked through it, and are very pleased with the results.

There have been technological problems, also, that have exceeded our expectations, particularly with that "last mile" - that last mile of local interconnection seems a very long one! There have also been problems with the teleconferencing bridges in both earth stations, and in the satellite, and with the end user equipment. One of things we have observed is that one can get brilliant engineering expertise to deal with almost everything except end-user equipment. There are very few engineering consultants who even know what equipment is available, let alone who really understand the problems of interfacing end-user equipment with the other elements of the system. We've also learned that when you diffuse technical responsibility between a supplier and a local ministry, you the get the problems you might expect - no one is quite fully responsible. And yet, there is a tradeoff, because technology transfer means that local organizations need to have the experience of carrying key activities out themselves. We've also relearned, as any of you in technology have learned, that no equipment arrives ready to operate. No equipment. So, it has taken a long time, particularly with the distances between suppliers and users to bring these systems up to par.

The more positive lessons have been numerous. Telecommunications authorities and social science <u>can</u> plan together! There <u>are</u> methodologies for interactive distance teaching that are readily transferable from society to society. These services are so popular that time on the circuits becomes a new kind of problem. Innovative uses are emerging that were not planned. And the technical problems are, in fact, soluble with enough patience.

As several nations begin to use their enormous new satellite capabilities, we will avidly look forward to the greater sharing of experience. I believe India's experience with INSAT, for example, will be absolutely vital and pivotal with regard to the rest of the world, and we know that experience is being shared generously. As we look to the future we are also expecting, of course, further cost reductions. They have taken longer than we had thought, but they are close.

I would say that we are looking to the future with a great deal of optimism, fundamentally because user acceptance has been so intensely positive. People are learning from these systems, and administrators are seeing new possibilities. During the next year, the cooperating countries and experts will be reporting extensively on their experiences. We look forward to sharing those, and to seeing other nations begin to use these powerful capacities to spur the development of their peoples.

THE ALASKAN SATELLITE EXPERIENCE: LESSONS FOR THE DEVELOPING WORLD

by

Walter B. Parker

President, Parker Associates

HISTORY

Telecommunications development in Alaska has always depended on government support and stimulation for research and for system development. Since the purchase of Alaska from Russia by the United States, it has been an area where innovative techniques have been pioneered. Even during the Russian period, the area was the center of the effort to build a telegraph line from the United States to Europe, an effort that was cancelled upon the successful allying of the Atlantic cable. In 1903, the first radio-telegraph link ever utilized was developed by the U.S. Army in Alaska to replace a submarine cable destroyed by ice action. This was the period when Alaska communications were under military control, a period that lasted until 1971, when the Alaska Communications System was sold to RCA. Even after the sale, the military was still the largest telecommunications operator and user in the state.

Since the system bought by RCA had been largely developed to meet governmental and military needs, one of the requirements of the sale was that telephone service would be rapidly extended to 143 communities that were completely without telephone service in the state. RCA agreed to this condition and planned to accomplish it through microwave extensions from regional communications hubs. These hubs were served by a mixture of microwave, troposcatter and satellite channels. Satellite service was based upon a system of 30 meter and 10 meter earth stations. These stations were too costly to consider installation in the small villages that lacked telephone service.

At the same time as RCA purchased the Alaska Communication System, the Applied Technology Satellite (ATS) experiments began in Alaska. Experiments in distant delivery of health services and education services utilizing ATS-1 were begun in 1971 at 26 locations. This experiment was limited to audio and data transmission, since the video channel on ATS-1 had failed. In a way, the video failure was a benefit since the experimenters were forced to learn how much could be accomplished through audio ultilization only. The experiment also proved the utility of small 3 and 4 meter earth stations and also the ability of these small stations to access the satellite from high latitudes.

The ATS-6 experiment was conducted in 1975 and focussed on video applications to some 16 communities. Again there were major health and education components to the experiment. The experiment proved that television could be received through small earth stations in Alaska. The health experiments opened new horizons for what was possible in providing medical service in remote communities, a host of expectations being created that are not fulfilled yet today for reasons which will be addressed later. The education experiments were less satisfactory and did not add substantially to the store of knowledge gained from three years of ATS-1 educational experimentation. The reason for this was that the experiment was simply too short to affect behavioral change of either teachers or students. It did, however, leave the rural people with the knowledge that television could be provided to their communities with existing technology and increased expectations greatly in that area.

The net result of the two ATS experiments was to create a group of people in Alaska, primarily in government and the universities, who were knowledgeable about the potential of satellite communications in providing service throughout a state that reaches 2,500 miles (4,200 kilometers) east and west and 1,100 miles (1,800 kilometers) north and south. A state with a population then in 1975 of some 320,000 that has now increased to 450,000. A state also with great climatic and geographical barriers to easy access through transportation. In short, a state whose need for a reliable, effective and affordable communications systems was obvious to even the most obtuse policy maker.

IMPLEMENTATION OF THE SATELLITE BASED SYSTEM

In December 1974, a new administration entered office in Alaska. The evaluator of the ATS-1 educational project became Commissioner of Highways, a cabinet post in Alaska at that time with the largest capital budget and the largest work force. The Director of the Alaska Educational Broadcasting Commission, who had been active in ATS-1 and ATS-6 experiments, became the new governor's Director of Telecommunications, a post in the Office of the Governor. This created the nucleus of a strong cadre with experience in the then new small earth station technology that had access to the governor and the legislature on a regular basis.

By the end of 1974, there was general dissatisfaction among the residents of rural Alaska about the rate at which RCA through its Alaskan subsidiary, Alascom, was providing telephone service to the 143 locations required in the agreement of sale. This feeling was reflected strongly in the Alaska legislature among both urban and rural legislators. In December 1974, the Commissioner of Highways advised the Governor of the problem and requested permission to set up a special task force to resolve it. He also requested $10,000 from the contingency fund for extra expenses and engineering consultants.

Work sessions and hearings were held with Alascom and RCA headquarters employees that were generally unproductive. It was decided to pursue a state owned system of small earth stations for rural services. Engineering advice was sought from the Stanford University Department of Electrical Engineering, a former Hughes engineer, and a University of Alaska engineer who was on loan from his role as a legislative consultant. Systems analysis support was supplied by the Stanford Institute for Communication Research, which had been the evaluator of the ATS-1 health experiment, and the University of Alaska.

Based upon the experience gained in the ATS experiments, a prototype small earth station was built from components available on the market in large part, with some special orders. This station's prototype had been designed by the University of Alaska Department of Engineering under a contract to the U.S. Public Health Service. More meetings were held with RCA and Alascom, and RCA agreed to modify their F1 and F2 satellites to provide more amplifier gain through the satellite and to adjust their antenna feed-horn so as to increase the radiated power directed towards Alaska for 6 of the 24 transponders on board each satellite.(1) By this decision, they agreed to the small earth station as the basic means of rural telecommunications delivery. Further meetings between the state task force and RCA/Alascom finalized the request for procurements which were developed to purchase earth station components. Monies were supplied by a $5 million appropriation from the Alaska legislature to purchase 100 small earth stations. After this action by the legislature, an agreement between the state and RCA/Alascom was laboriously reached under which the state owned the earth stations and Alascom maintained and operated them. All of this action took place between January 1, 1975, and June 1 of the same year.

At the same time, the state was challenging RCA/Alascom before the Federal Communications Commission (FCC) on the relationship between RCA Satcom, the satellite owner, and RCA Alascom, the satellite users. The aim of the state effort was to insure that the FCC's "open skies" policy on satellite leasing was used in the Alaska situation also to guarantee competitive pricing. It must be remembered that deregulation was just beginning at this time and that the forecast for satellite availability in the 1980s was for a highly competitive atmosphere, in short a "buyers market", rather than the situation which we have now.

In any case, the State did purchase the earth stations and Alascom did install them through 1977. The stations were designed to provide a minimum of two circuits, one for telephone and the other for emergency medical service to be operated by the Public Health Service. Expansion to eight circuits was possible through insertion of circuit modules.

In villages without a local telephone exchange, a single instrument was located in a central place as a public telephone. It was hoped that local telephone companies would move rapidly to provide service at these places but these hopes proved unfounded. The Alaska Public Utilities Commission did not make a major endeavor to insure that local services were provided at all outlets, and it was not until the past year that all villages having small earth stations acquired local exchanges. This was due to Alascom, which was purchased by Pacific Power and Light in 1978, taking the initiative. Pacific Power and Light set up a holding company, Pacific Telephone, which became the owner of Alascom. Pacific Telephone purchased Sitka Telephone, and that company made application to provide local services to the remaining villages without local exchanges. There were some 40 remaining, and the Sitka Telephone application spurred some other locals to make applications.

The net result is that finally all Alaska villages with satellite service now have local exchange capability some eight years after the small earth stations were installed. The policy failure was that the local exchanges were not made a part of the initial push in 1975 to provide rural telephone service, but were left to evolve without strong stimulation from the state government. Evolution took far too long.

IMPLEMENTATION OF PUBLIC TELECOMMUNICATIONS SYSTEM

In the summer of 1975, one of the engineers on the state team that designed the small earth stations was appointed Director of an expanded Office of Telecommunications, which remained a part of the Office of the Governor located in the capital, Juneau. His predecessor moved to the Alaska Public Utilities Commission and continued to be heavily involved in regulatory matters affecting telecommunications development. The special task force was disbanded, having achieved its major objective of switching rural telecommunications to a satellite based system. The Alaska Legislature had not lost its enthusiasm for telecommunications, however, and in 1976 appropriated $1.5 million for a rural television demonstration utilizing the small earth stations that were still being installed. A transponder was available on Satcom F2, and the Office of the Governor concurred with the legislative desire. The demonstration provided private broadcasters in Alaska's cities and towns with same day network programs for the first time, with news and sports receiving first priority. In exchange, the broadcasters provided programs to the statewide rural network. Thus the statewide system received programs from all three major U.S. networks plus public broadcasting material. Program selection was made by a committee of rural residents. This uneasy alliance between the state and the private broadcasters has now existed for eight years, and the demonstration program is an institution serving all villages having small earth stations.

The system is now showing signs of wear. Private broadcasters now have their own real-time television feeds from their networks in some cases, and only a few now utilize the state transponder. The other development is that many villages have installed their own antennas and cable distribution systems; thus the state channel is competing with 12 or more channels of television at many locations. There is discussion about eliminating the state system, and this possibility is under evaluation now.

Local distribution of the state system was by mini-TV transmitters with a range of about one mile. Those villages not cabled and persons in cabled villages who are not on the cable still receive their only television from the state channel. It is not known yet how powerful a constituency this is.

The next step in public system development was also initiated by the legislature. In 1977, the Legislative Teleconferencing Network (LTN) was created with service at six sites in the major cities authorized. The system utilized audio conferencing initially, but video conferencing was funded in 1978 on an occasional use basis. Legislators have not responded enthusiastically to video conferencing because they find it cumbersome and time consuming. They like audio conferencing because they can drop in and out of the conference room without being observed by the listening audience, whereas with video conferencing the need to remain on camera is strong, since no one wants to be observed as absent by his constituents. In essence, video conferencing does not suit the legislative way of doing business as well as audio.

The system was expanded to 11 sites in 1978 and to 15 in 1980. In 1982, 40 villages were added with funded site coordinators. The LTN sites can exchange print materials in most cases. The LTN system is mature and well staffed. It has made possible participation in legislative hearings by persons from the far corners of the state. There is no evidence that it has particularly improved the quality of Alaskan politics, which are somewhat too tempestuous at the moment for any quality indicators to be established.

As its next step in telecommunications development, the legislature in 1979 requested the University of Alaska and the state Department of Education to undertake an assessment of the need for a statewide educational television system. The group undertaking the assessment included two from the original special task force of 1975, the former Commissioner of Highways and the University of Alaska engineer.

The assessment involved the state's 53 school districts and some 30 university units. The recommendation to the legislature was for a single television channel to be jointly used by the school districts and the university but, more important, for an extensive audio conferencing network to serve all schools in the state. The initial priority was for rural schools, but the long range intention was and is to include urban schools in the network also. Computer networks were left to future development, since the university was not using its computer network intensively for instruction, and the Department of Education had an ongoing demonstration on computer assisted instruction using stand alone microcomputers which was operating successfully without networking at that time.

The legislature accepted the recommendations of the assessment and funded half of the system for rural areas in the FY81 budget and the other half in FY82. Another 100 of each were procured in the next budget cycle, and the system now reaches over 200 locations. The typical broadcast day is 16 hours, with broadcasting to primary and secondary schools through the day and to the university in the evening.

The audio conferencing system was set up with four Darome bridges operating in a central location at Anchorage. Thus, 80 sites can be teleconferenced simultaneously or any combination up to 80. A regional bridge based at Kotzebue was funded in the last legislature in 1984, bringing the system capacity now to 100 sites.

Audio conferencing has proven extremely popular as it was expected to be. Over 200 courses have been offered over the network, and it is used intensively for tutorial and professional enhancement also.

Utilization of television is considered high, but a thorough evaluation is needed of exactly how television is being utilized. It has not been possible to fulfill demands for regional and culturally relevant materials because educational programs have been consistently underfunded from the first. This is not particularly the fault of the legislature, since no strong requests have been made by either the university or the Department of Education for a budget that would provide programming at the level that the network clients indicated is desired. Besides formal educational materials, there is a strong need for adult education for Alaska Natives and others in a wide variety of areas such as business management, resource development, land planning and environmental protection. The original conception of the network was that it would be utilized extensively for this type of education and that another channel would be sought if necessary.

There is expressed concern from some rural legislators that the system, known as Learn/Alaska, is not working well, and an interim evaluation has been funded but not implemented as yet. Conversations with many rural teachers and written communications express a reasonable degree of satisfaction. Hopefully, the planned evaluation will reconcile this dichotomy of views.

The original premise upon which the Learn/Alaska system was instituted has not changed. The State of Alaska is required to offer a high school education in every village by law. This means most rural schools have from two to four teachers attempting to teach twelve grades. The major role of Learn/Alaska is to provide backup to these small teaching staffs so they can provide what is accepted as a normal high school education in the United States. The rural university units have the same general problem, and the university hopes to expand its programs in rural areas through utilization of Learn/Alaska by using faculty on the larger urban campuses. So far, this development has been spotty, but the Anchorage and Fairbanks schools of nursing have shown it can be done, and other departments and colleges are beginning to participate.

Finally, we come to the oldest public system in Alaska, that operated by the U.S. Public Health Services to provide contact between the doctors in regional hospitals and the health aides at 120 sites across Alaska. Health aides are para-professionals who are selected by their villages and provided ongoing training by the PHS. They utilize both conference circuits and telephones depending upon the need for security of communications. They system has been utilizing satellite communications since 1971, and utilized HF before that. Federal funding has not been available to expand into some of the systems tested by the ATS-6 experiments. Exchange of visual materials via slow-scan is being funded by the North Slope Borough for its health aides, and they will also be able to receive print material.

Although the North Slope Borough serves only seven of the 120 villages with health aides, evaluation of their new systems will provide guidance on the utility of expanding such services to the rest of the health aides, something that has been needed since the ATS-6 experiment ended. It has not been done because the Public Health Service could not secure federal funding and the state has been reluctant to assume what it regards as federal responsibilities.

These four major public networks have made possible access to a wide spectrum of information and entertainment throughout Alaska. The utility of the health system is well accepted by most obervers; that of the educational system less well since it is much newer and has not been subject to major analysis as yet. The utility of the legislative system is accepted, but because it is a branch of the legislature, it has received little analysis, except for the televising of legislative sessions which has been underway since 1978 and which has engendered a great deal of controversy. Finally, the utility of the entertainment channel is accepted because of continued strong public support in rural areas for its continuation, a factor which, as previously discussed, may be subject to change.

The analysis of televising legislative session (2) indicated that the programming was highly desired by those interested in government and politics. It also indicated that this group constituted a small part of the total population in both urban and rural Alaska. The conclusion was that though this group was small, it was in the best interest of the state to keep activists informed, since new recruits to the political and governmental process will come from their ranks.

IMPLEMENTATION OF PRIVATE SECTOR NETWORKS

It is not surprising to find that those Alaskan entrepeneurs that are successful in their fields have also been pioneers in utilizing telecommunications networks. The state's major wholesaler, J. B. Gottstein, Inc., has operated a network for several years to some 85 small village stores that allows the store managers to update their inventory requirements daily into the central computer. Combined with more reliable and more frequent air transportation, this has enabled the most remote stores to operate on short shelf life inventories, rather than storing goods for up to a year as was past practice.

ARCO has developed ARCONET, a sophisticated system that links their Arctic operations at Prudhoe Bay with their headquarters in Anchorage, Los Angeles, Dallas and Denver via teleconferencing, data displays and electronic mail. This is probably one of the most sophisticated integrated systems extant.

There are many private computer networks that are similar to those springing up elsewhere. Engineering firms with several locations throughout the state are rapidly expanding their capabilities to audio conference and exchange print material.

Alascom is offering ALASKANET which makes the conventional data bases available in North America available to Alaskans. It has not been in operation long enough for observers to have any sense of its utility. Many Alaskans were already using the Source, Compuserve and other data bases in any case.

IMPLEMENTATION OF PUBLIC TELECOMMUNICATIONS INSTITUTIONS

There was no central responsibility for telecommunications policy in the Alaska state government until 1971. An Office of Telecommunications Policy was then set up in the Office of the Governor in order to give it policy making status and authority to coordinate disparate public telecommunications efforts in several state agencies and commissions. This office was expanded in 1975 and maintained as a major force in telecommunications policy making until it was abruptly eliminated in 1979 because of political pressure from the private sector. Its functions for coordination and operation were transferred to the Department of Transportation and Public Facilities, where it became purely a support agency with no policy function.

The Office was revived in 1980 as part of the Learn/Alaska development surge in the legislature. Two divisions were created, the Division of Telecommunications Systems and the Division of Telecommunications Services. They were placed in the Department of Administration under a Deputy Commissioner for Telecommunications, primarily because no other more suitable home could be found. The governor did not want them back in his office, and the legislature would not consider an independent agency. Two Deputy Commissioners have come and gone, and nothing much has happened. The last Deputy resigned under pressure in the summer of 1984, and has not yet been replaced. The operation appears to be almost completely moribund insofar as policy initiatives or new systems development are concerned. Despite the critical situation in which telecommunications deregulation places the State of Alaska, the present state presence is probably the weakest it has been since 1969, since during that period, though there was no institution, the governor and his attorney general took a good deal of interest.

The State of Alaska's experience with telecommunications institutions is little diffferent from that of the federal government or other states. The blend of private systems that operate telecommunications in the United States does not respond well to strong policy direction, even though its recent experience with deregulation effected through the federal Department of Justice should give it some pause for concern.

The great reliance upon telecommunications in Alaska and the great public investments that have already been made give it a special reason for providing strong public policy leadership. This leadership has come in the past from the legislature largely, and reflects the general public demand for telecommunications services. When legislative pressure has been matched by some response from the administration, things have happened quickly. Now the important issues affecting telecommunications systems are largely being left to the state regulatory process through the Public Utilities Commission, an agency somewhat hampered in policy making by its regulatory constraints, and to the courts. It is too bad that having come so far in telecommunications development, a hiatus has occurred at this most critical period.

The other lesson that can be learned from institutional development in Alaska is that no organizational structure can rise above the individuals involved, but that organizational structure can keep individuals capable of leadership from being effective. Most of the group who effected the changes of 1975 and the development of the public networks are still active in the state, but the conditions that bring about strong public policy development are not present -- mainly a strong focus in a leadership position in the state administration to which they can adhere.

CONCLUSION: LESSONS FOR THE DEVELOPING WORLD

Hopefully, most of the following conclusions are reasonably well substantiated or indicated in this paper. Space has precluded as full a development as would be required for complete treatment. Each conclusion could easily justify an article of this length and a book in some cases.

The first and most important lesson is that the provision of reliable telephone service has yielded the greatest social and economic returns. This has made possible the delivery of a wide range of audio, data and print services in a host of public and private applications. These services have proven thus far to be the most cost effective and the most neutral to cultural bias. Our great policy failure in this area in Alaska, was our failure to provide local distribution systems as a part of the original earth station program.

Television services are attractive enough politically to generate public support, but they should not be implemented at the cost of basic services. The cost of television program development makes it the most culturally biased since even relatively well-to-do states, such as Alaska, will not fund a high level of program development that has local relevance. Commitment should be secured to development of local materials when the system is funded, or the alternative accepted of utilizing material developed for other societies. The failure to fund the development of culturally relevant material on a timely basis must also count as a failure of policy in the Alaska sense. In this sense, locally relevant material does not apply only to material developed for Alaska Natives -- Inuit (Eskimos), Indians and Aleuts -- but also for the 87 percent of the population that has moved to Alaska from other parts of the United States, East Asia, Europe and Latin

America. Television must be a primary device for acculturating these immigrants to Alaskan society's past history and present needs.

Support of para-professionals working in rural areas in health, education and other areas can benefit from professional support delivered by telecommunications. Most important is audio, then print, then video. The most critical area that is usually overlooked in funding is the training of the providers in telecommunications support systems. Within the university in Alaska and within federal and state agencies, there is still a good deal of opposition to distant support systems. Traditionalists can only be won over by incorporating them in the system, ultimately. The way to do this is through early training programs that force them to confront their biases. The record in Alaska is mixed on this. Learn/Alaska devoted substantial portions of its early budgets on teaching audio conferencing techniques to rural teachers. As a result, a strong cadre of support was formed early in the system development.

Centralized technical support agencies should be viewed with care. There seems to be a tendency of such agencies to regard their support mission as more important then the missions they are funded to support. Except where economies of scale demand such centralization, it is wise to fund such support systems within the agencies operating the systems, whether those agencies be schools, universities, hospitals or transportation agencies. This comment applies to special networks and not to the general telephone system.

Once the demonstration or start up period is finished, agencies providing delivery of services must be required to justify and support those services in their own budgets, rather than from a special telecommunications budget. Otherwise, budget coordination becomes cumbersome and gaps are left too easily.

If a centralized policy development institution is created, it must have more political power and access than the organization sfor which it is designed to provide policy direction for. This is true, whether the delivery system is privately owned or publicly owned. Power in this sense does not imply arbitrary direction, but rather the power to convene disparate users and providers in a continuing consensus-achieving and planning effort. Planning in this sense is strictly policy planning and not implementation planning which should be left, in most cases, to those who will implement.

The implementation of technical systems should begin with the needs of users and work upward to the satellite.

Too often, the high cost items are funded and those low cost items, such as video tape recorders, microcomputers, audio recorders and data terminals that will provide the system flexibility and maximum utility, are overlooked or strictly rationed.

The above lessons reflect the Alaska experience only. They should be used with care in different political, social and economic situations from those pertaining in Alaska, but there are hopefully some findings that are relatively universal to the human experience as it seeks its new future in a rapidly developing age of information.

REFERENCES

1. Merritt, Robert P. "Alaska Telecommunications" in Telecommunication in Alaska. Papers in support of the Alaska Case Study Presentation to the 1982 Pacific Telecommunications Conference, Edited by Robert Walp. Pacific Telecommunications Council, Honolulu, 1982.

2. Parker, Walter B. "Opinions for Televising the Alaska Legislature" Unpublished report prepared for the Alaska Legislature at the direction of the Alaska State Library System, May, 1982. Copies available from author at 3724 Campbell Airstrip Road, Anchorage, Alaska 99504.

DISCUSSION

QUESTION:

Has Mexico decided on the video standard it will utilize?

MR. SEGOVIA:

The American video standard.

QUESTION:

What impact will the earth stations in Indonesia have for rural applications?

DR. BLOCK:

The cost of the Indonesian earth station is considerably lower than former ones, at $90,000. This is still higher than it needed to be, because it had to be designed for solar power in order for this kind of communication technology to really make a difference in the Third World, costs have to come down to under $25,000. Some people believe that now, with sufficiently large orders of earth stations, the cost could be reduced to around $35,000.

QUESTION:

Please follow-up on earth stations, low cost stations, especially C-band stations for oil rigs, etc. How do you plan to get the benefits of satellite to some of the small villages in your countries and still get costs down?

DR. SRIRANGAN:

We are operating television in both C-band, and S-band; both re-broadcast quality. For community reception, 42-43 dB of noise we think for rural application is quite adequate. At present, cost as installed with the associated television set is about $25,000 per station. With respect to telephony, we do not have a low-cost terminal. The smallest we make is 4.5 meters, although for emergency services (jeep or aircraft, etc.) we have smaller terminals with some trade-off in spectrum use.

We are addressing issues of reducing costs. The main problem has been that the costs of production is related to quantity of production so we have not been able to realize savings on that scale. Nevertheless, with the initial exploitation of C-band transponders there has been tremendous explosion in demand; projected demand is already up to 300-400 such terminals. We are hoping that costs be reduced in future to $5,000 to $10,000 each.

MR. SEGOVIA:

We are trying to operate in Ku-Band; we hope to have smaller antennas. Actually there has been an ongoing program with which I am not too familiar -- it has not been decided how and what technology will be used for the rural areas. Testing of Mexican-made solar panels and equipment, etc. is one of the pilot projects.

QUESTION:

Could you tell us more about use of microcomputers with the system in Indonesia? I have been working with a group to propose addition of microcomputers to a statwide telephone network in Texas, and we think we have seen certain advantages to the micro. I would like to know what you think might occur in experimentation.

DR. BLOCK:

We haven't done much yet. It has been a recent addition to the Caribbean project and the Indonesian project. It has been experienced longer, as Heather Hudson knows, in the University of South Pacific where "Apples" were introduced and multiplied. They are being used successfully for electronic mail, which turns out to be terrifically useful. In the South Pacific, the alternative is the longboat which takes weeks. In addition, the ability to communicate without having someone manning the station is very valuable. I think it has enormous potential for communication, for transmitting hard-copy data in support of education, certainly for all sorts of administrative purposes.

One of the things we found is a strong requirement to see something in print, because it is a more official statement of policy or direction than a verbal statement. I failed to mention that we are also using facsimile in the initial project, which has proven useful as well.

QUESTION:

You seem to indicated that the development of the satellite program in India is somewhat of an overall attempt to develop capability in the area of telecommunications. To what extent do you relate that to very concrete, specific state policies in India and how does that relate to an overall development of a microelectronics industry in India itself?

DR. SRIRANGAN:

Basically, we consider satellite as only one of the ingredients of a telecommunications structure, although a very important one. We continue to place a large reliance on the question of what has application for a country of our size, the population, and the kind of traffic we have. There are many angles to the issue of policy. We recognize satellite as an important element of the policy framework, but as of now, very largely, the entire telecommunications is in government hands, although there are winds of change blowing through. And that about sums it up.

PANEL 4: TELECOMMUNICATIONS REQUIREMENTS IN DEVELOPING COUNTRIES

The purpose of this panel was to examine the broader issues of telecommunications requirements for developing countries, and the steps which must be taken for developing countries to obtain telecommunications facilities including satellites, and to use them for development.

The first speaker, Dr. Heather Hudson, is an Associate Professor in the Department of Radio-Television-Film at the University of Texas at Austin, where she teaches and does research on applications of new technologies, and telecommunications policy. Her research interests include the role of telecommunications in development and applications of new technologies, particularly satellites, for social services and rural development. She has planned and evaluated satellite projects in northern Canada and Alaska and in numerous developing countries, and has been a consultant to the Independent Commission for Worldwide Telecommunications Development (the Maitland Commission) and a member of the U.S. Advisory Committees for the 1979 WARC and the 1985 Space WARC. Her paper reviewed some of the major findings on the role of telecommunications in development, and outlined some of the risks which have acted as disincentives to investment in Third World telecommunications.

The second speaker, Dr. Edwin Parker, is Vice President of Equatorial Communications Company in Mountain View, California. Equatorial manufactures micro-earth stations for transmission and reception of data transmitted by satellite using spread spectrum techniques. It is also a common carrier, with its own satellite capacity for data transmission. Equatorial facilities are used for transmission of news, weather data, agricultural information, and stock and commodity prices. Edwin Parker was formerly Professor of Communication at Stanford University, where he directed numerous research projects in the planning and evaluation of applications of new technologies including computer networks and satellites for libraries, rural health services, and rural development.

The third speaker was Dr. Joao Carlos Albernaz, Secretary of Informatics in the Brazilian Ministry of Communications. He was previously Secretary for Planning and Technology. Dr. Albernaz has represented Brazil at many international policy conferences, and headed the Brazilian delegation to the 1977 WARC on Broadcast Satellite Services. Dr. Albernaz' paper outlined the history of satellites communications in Brazil, and plans for the domestic BRASILSAT system.

The final speaker on this panel was Mr. Richard Stern, Chief of the Telecommunications, Electronics, and New Technology Division of the World Bank. Mr. Stern was trained in economics at the Institute of Development Studies of the University of Sussex. He worked for the United Nations in Ethiopia, was the division chief for Indonesia in the World Bank, and was the economist for the World Bank for East Africa. Mr. Stern's presentation outlined the developing country context and the financial and organizational issues that must be addressed in planning telecommunications expansion in developing countries.

RISKS AND REWARDS: WHY HAVEN'T TELEPHONES REACHED MORE VILLAGES?

by

Dr. Heather E. Hudson

Associate Professor

Department of Radio-TV-Film, University of Texas at Austin

INTRODUCTION

Advances in communications technology--from satellites to solar powered two-way radios--now make it possible to extend reliable communications to any village or camp, whether in the desert or the jungle, or on a remote island. But progress in taking advantage of these technological advances to meet the needs of rural people in the developing world has been painfully slow. Telecommunications has been considered a luxury to be provided only after all the other investments in water, electrification, and roads, etc. have been made--and after all the demand for telephone services in the cities has been met.

Yet, telecommunications should be considered a vital component in the development process--a complement to other development investments, that can improve productivity and efficiency of rural agriculture, industry, and social services, and can improve the quality of life in developing regions.

Telecommunications was virtually ignored by development theorists and planners until the past decade. However, recent studies, many of them sponsored by the ITU and the OECD Development

Centre, have shed considerable light on the ways and extent to which telecommunications contribute to development. While more research will be necessary to refine our knowledge, several conclusions can now be put forward:

1. The indirect benefits of telecommunications generally greatly exceed the revenues generated by the telecommunications network;

2. The indirect benefits of telecommunications may contribute significantly to economic activities and delivery of services in developing regions;

3. Investment in telecommunications contributes to national economic growth;

4. Benefits of telecommunications are related to distances between communities of interest and telephone density (telephones per 100 people), so that benefits of investment in telecommunications may be greatest in rural and remote areas;

5. Telecommunications can be considered a complement in the development process; i.e. other pre-conditions and co-conditions must exist for maximum developmental benefits of telecommunications to be achieved;

6. There are intangible benefits of telecommunications which, while difficult to measure, can contribute to the development process.

INDIRECT BENEFITS: TELECOMMUNICATIONS AS A PUBLIC GOOD

The indirect benefits of telecommunications are related to the function of conveying information. Information is critical for any development activities--for administration and management, trade, consultation, etc. Information also has unusual economic properties in that it can be shared without being transferred and that its benefits may extend to others besides those directly involved in the information transaction. For these reasons, information, and by extension, telecommunications as a means of instantaneously conveying information, may be considered public goods. The benefits of a telecommunications call, for example, may accrue both to the caller and the person called, but also to others who are not involved in the information transaction.

But even more important in terms of development is the fact that the society as a whole will benefit from these uses of telecommunications. Access to a physician can reduce mortality rates, and allow more people to be treated adequately by village health workers without having to be transferred to a hospital. Consultation with agronomists and veterinarians can improve crop yields and livestock production. Use of telecommunications to order supplies can result in faster supplying of rural communities and reduction of wasted trips by villagers and shippers.

Yet planners do not generally take these "indirect" benefits into consideration in determining how much to invest in telecommunications and where to install new facilities. These decisions are more likely to be based on the anticipated revenue to be derived from the telecommunications network. However, the revenues are not likely to reflect the full benefits to the society of the telecommunications system.

Telecommunications networks could be considered basic infrastructure like roads, water mains, and electrical power grids. Planners do not expect these services to make money directly, but rather to facilitate the development process. However, unlike other utilities, a telephone network can generate its own revenues, and thereby become a profitable investment for a country. Yet this potential profitability is almost an "Achilles heel" from a development perspective, in that planners tend not to look beyond the revenues to determine what greater benefits might be gained from the telecommunications investment. Furthermore, they may expect the revenues from each part of the network to cover their costs, so that rural areas may be considered inherently unprofitable, and therefore not deserving of investment of scarce resources until urban needs are satisfied. But this approach overlooks the important developmental value of the indirect benefits of telecommunications.

Three important conclusions can be derived from the public good nature of telecommunications:

* there is likely to be a surplus of benefits over costs to the user of a telecommunications system;
* these benefits may accrue to both parties in the transaction and to others in the society as well;
* revenues from calls may not reflect the indirect benefits derived; or from a requirements perspective, demand as measured by expected traffic and revenues may not adequately reflect the need for telecommunications.

DEVELOPMENTAL BENEFITS OF TELECOMMUNICATIONS

Access to information is key to many development activities, including agriculture, industry, shipping, education, health and social services. The ability to communicate instantaneously can facilitate the development process in three major ways--by improving:

* <u>efficiency</u>, or the ratio of output to cost;
* <u>effectiveness</u>, or the extent to which development goals are achieved;
* <u>equity</u>, or distribution of development benefits throughout the society.

The developmental benefits of telecommunications for economic activities and social services have been identified in several case studies sponsored by the ITU in collaboration with the OECD Development Centre and by the World Bank (1,2) as well as by other development agencies (see 3). Macrolevel economic studies have examined the contribution of telecommunications to economic growth, while microlevel analyses and case studies have emphasized the ratio of direct and indirect benefits to costs of telecommunications investments.

Among the benefits of telecommunications for improving efficiency and productivity are the following:

* Price information: Suppliers such as farmers and fishermen can compare prices in various markets, allowing them to get the highest prices for their produce, to eliminate dependency on local middlemen, and/or to modify their products (types of crops raised or fish caught, etc.) to respond to market demand.
* Reduction of downtime: Timely ordering of spare parts can reduce time lost due to broken pumps, tractors, etc.
* Reduction of inventory: Businesses can reduce the inventories they need to keep on hand if replacements can be ordered and delivered as needed.
* Timely delivery of products to market: Contact between producers and shippers to arrange scheduling for delivery of products to market can result in reduced spoilage (e.g. of fruit or fish) and higher prices for produce.
* Reduction of travel costs: In some circumstances, telecommunications may be substituted for travel, resulting in significant savings in personnel time and travel costs.
* Energy savings: Telecommunications can be used to maximize the efficiency of shipping so that trips are not wasted and consumption of fuel is minimized.

* Decentralization: Availability of telecommunications can help to attract industries to rural areas, and allow decentralization of economic activities away from major urban areas.

Examples of these benefits can be found in many parts of the developing world. In Sri Lanka, small farmers were able to use newly installed rural telephones to obtain prices of coconut, fruit, and other produce in Colombo. As a result, instead of getting 50 to 60 percent of the Colombo price for their products, they were able to get 80 to 90 percent of the urban price.(4) In Alaska, once a satellite earth station for telephone service was installed, a fish packing plant in the Aleutian islands was able to fill orders from its headquarters fasters and to change the type of fish caught in response to changes in New York prices.(5) And Indian trappers in northern Canada may now call to compare prices offered for their furs by co-operatives and by fur auction houses in the cities rather than selling only to the general store in the village.(6)

Similar effects can be achieved at the national level. Since the installation of INTELSAT satellite earth stations for international communications in the South Pacific, government officials have been able to save money on procurement by sending out tenders and receiving bids by telex from international suppliers rather than placing standing orders with one supplier.(7)

A study of businesses in Kenya estimated that the losses incurred as a result of poor telecommunications were on the average

110 times higher than the total cost of providing adequate telephone service, and amounted to an average of 5 percent of total turnover.(8)

Studies by researchers at the Moscow Electrotechnical Institute of Communication also showed substantial industrial benefits from telecommunications use. Each "communication unit" of 1000 long distance calls resulted in a saving of 3700 personnel hours or 36,000 rubles in production costs. The researchers also estimated the income-generating effects of telecommunications. Together these benefits of telecommunications use were 4.3 times higher than the cost of the telecommunications services.(9)

Improved coordination of transportation is important in the marketing of perishable products. In the Cook Islands, a two-way radio network is used by agricultural officers to notify the shipping agency about the amount of fresh fruit ready to be picked up from each island during the week. The shipper then sets the schedule, and notifies the farmers so that they will have the fruit ready for the ship's arrival. Without this information, farmers risk spoilage if the fruit is picked too soon, and the shipper risks major delays if the boat must wait at each port for the fruit to be delivered.(10)

Telecommunications use can also reduce the need for travel. In India, the benefit to villagers of using long distance public telephones was about 5 times the cost of the call, taking into consideration the cost of bus fare and of the time lost from work in travelling to town to deliver a message.(11) In Egypt, researchers at Cairo University calculated that benefits to villagers of using rural public telephones could amount to more than 85 times the cost of the call, taking into consideration such factors as the value of production that would be lost if equipment were to remain idle while waiting for spare parts, as well as the savings in travel time and costs.(12)

Related to travel savings is the ability to make more efficient use of necessary transportation. This benefit could be extremely important for developing countries dependent on imported fossil fuels which are a major drain on their foreign exchange. Transportation accounts for 10 to 20 percent of the total energy consumed in the low income developing countries. A study carried out for the ITU estimated that through better telecommunications, it should be possible for the oil-importing developing countries to save about $18 billion in oil imports.(13)

Similarly, education, health, and other social services may also benefit directly and indirectly from telecommunications. Several of the benefits enumerated above, including savings in travel and energy, and improved management of decentralized activities, also apply to social services. Indirect benefits of telecommunications may also include effectiveness and equitable access to services.

Quality and effectiveness of rural services may be improved if rural workers can consult with physicians in cities or regional centers; agricultural extension workers able to obtain information directly from experts to meet farmers' needs; correspondence students able to interact with their instructors. Economic benefits may be derived from these information exchanges, such as reduction in number of patient evaluations, improved crop yields, and reduced student drop-out and failure rates.

INVESTMENT IF TELECOMMUNICATIONS CONTRIBUTES TO NATIONAL ECONOMIC GROWTH

Numerous studies have demonstrated a strong positive correlation between telephone density and economic development measured by GNP per capita or a similar statistic. However, this simple correlation did not explain what appears to be a chicken and egg relationship: does investment in telecommunications contribute to economic growth, or does growth lead to investment in telecommunications?

A model developed by Hardy using cross-lagged correlational techniques and time series data has shown a small but significant contribution of telecommunications investment (measured by telephone density) to economic growth (measured in GDP per capita).(14) This model can be used to predict the economic impact of telecommunications investment for regions with specified telephone densities. The rate of extension of telephone service can also be varied in the model.

The model shows that telecommunications leads or precedes economic development. However, economic development also contributes to telecommunications expansion, as more resources become available for investment in the sector. But now for the first time we have statistical verification of what appears intuitively obvious: that investment in telecommunications can contribute to economic growth.

The model has been used to predict the economic impact of installing satellite earth stations for telephone service to rural and remote communities, using telephone densities and economic data typical for rural regions of developing countries. It was estimated that each earth station with its associated telephones would contribute approximately $370,000 to the GNP over 10 years.(15)

WHO ARE TELECOMMUNICATIONS USERS?

As noted above, people who need to communicate quickly or frequently for their work include entrepreneurs, project managers, and health care providers. But rural residents may also use telecommunications facilities for many purposes. In Egypt, researchers found that better educated individuals were more likely to make calls to major cities and administrative centers, whereas those with little education tended to call only to nearby villages and towns.(16) However, the most important characteristic of telephone users is thirst for information. Village chiefs without formal education may use the telephone to talk to other chiefs.

Villagers who do not speak the national language or have limited education may rely on intermediaries to obtain information. They may to to a government official or village manager who in turn will use the telephone to obtain the information they need. Or they may call a regional or tribal organization whose staff will follow up to find the answers to the problem.

Thus although telephone users tend to be better educated and more involved in the market economy than non-users, literacy and "modernity" are not prerequisites for telecommunications use. Information seekers may be traditional people concerned about their families, their work, or problems in their community. They are likely to use whatever tools are available from two-way radios to satellite circuits to find the information they need.

BENEFITS ARE GREATER IN RURAL AND REMOTE AREAS

Information seekers who live in rural and remote areas tend to grasp immediately the benefits of telecommunications, because they know that the only alternative means of communicating quickly is through personal contact, which is likely to require a time-consuming and often expensive trip. In many parts of the developing world, villages are isolated for weeks during the rainy season, when flooded roads become impassable. The telephone or two-way radio becomes a lifeline not only for emergency assistance, but to keep up the contact necessary to administer government services and manage development activities and to reduce the sense of isolation.

Where telecommunications services are available, rural people often use them more heavily and spend more of their disposable income on telephone calls and telegrams than do city dwellers. In the Australian outback, "chatter channels" on two-way radios are busy all day long with messages in many aboriginal languages. In northern Canada and Alaska, Indians and Inuit (Eskimos) spend more than three times as much as their urban counterparts on long distance telephone calls, even though their average income is generally much lower than that of urban Canadians and Alaskans.

In these northern communities, telecommunications authorities have had to activate extra circuits in village satellite earth stations much sooner than they had anticipated because of the growth in telephone use. The number of long distance calls in some Indian villages in northern Canada increased by as much as 800 percent after satellite earth stations replaced high frequency radios. In Alaska, there was also tremendous growth in telephone use with the installation of small satellite earth stations in villages; another spurt of growth in traffic of up to 350 percent occurred when local telephone exchanges were installed in some villages, so that people could make long distance calls from their homes or offices rather than from a public telephone in a store or government office.(16)

Thus, in economic terms, benefits of telecommunications use are related to distance. Demand for telecommunications in rural areas is generally highly inelastic because of the lack of alternatives. The greater the distance from communities of interest, the greater the savings in travel costs and time from using telecommunications. Similarly, as noted above, benefits per telephone are likely to be greatest where telephone density is lowest. The implication is that the greatest payoff from telecommunications investment may be in rural and isolated areas.

ADVANCES IN TELECOMMUNICATIONS TECHNOLOGY

In recent years, advances in satellite technology have made it possible to extend reliable telecommunications to virtually everyone on earth, no matter how isolated. Via satellite, the cost of communication is independent of distance, as any two or more points can be linked directly without terrestrial lines or repeaters. Earth stations can be installed anywhere, so that telecommunications can be provided where needed, even in the remote areas which may need it most, rather than only as terrestrial networks are extended. Satellite earth stations are now small enough to put on a roof or even inside, aimed through a window. Similarly, new microprocessors have been used to reduce the size and cost of small switching systems and rural radio telephone networks.

Many of the technologies originally designed for use in industrialized countries may now be transferred with only minor modifications to serve the needs of the developing world--such as electronic PABX's and rural digital exchanges, small earth stations, spread spectrum transmission techniques, DAMA (Demand Assignment Multiple Access), cellular radio, and photovoltaics.

RISKS AND REWARDS

Why is it taking so long for the telephone to reach the village? Given the evidence of the contribution of telecommunications to the development process outlined above, and advances in technology with direct applicability to developing country needs, why are there still millions of villagers with no reliable communications available in their communities and millions more city dwellers with virtually no access to a telephone? The answer, I believe, must be seen in terms of relative risk. There is still considerable risk in providing telecommunications in the developing world, particularly in rural and remote areas. There is little recognition of the risk in not providing telecommunications. The risk in providing telecommunications can take many forms:

* the national planner risks his or her credibility by recommending investment in telecommunications rather than more "traditional" infrastructure and services;
* the planner risks prolonging complaints from vocal urban residents if they perceive rural telecommunications installations are being given priority over upgrading or expanding urban facilities;
* the minister of finance risks pledging scarce foreign exchange that may also be needed for other imports;
* the telecommunications authority risks a reduction in its rate of return if it extends its services into less populated areas;
* development agency officials risk their credibility

because they find telecommunications harder to justify than "basic human needs" programs emphasizing food and shelter;

* equipment suppliers risk payment delays or complications if there are foreign currency shortages;
* equipment suppliers risk time and money seeking markets in the developing world rather than concentrating on the more familiar and accessible markets at home.

Thus technological change and even a better understanding of the contribution of telecommunications to economic development are not sufficient to increase significantly the investment in developing country telecommunications. Some of these risks may be short term phenomena. For example, the booming telecommunications market in the industrialized world may peak, making foreign markets more attractive. The problem of foreign exchange may be less critical if oil prices decline, placing less of a strain on fuel importing nations. But the risks will not totally disappear. In order to stimulate investment in Third World telecommunications, it will be necessary to develop strategies that will both reduce risks and increase potential rewards.

Risk reduction can take many forms. It may include not only more definitive knowledge of the role of telecommunications in development, but also better efforts to communicate that knowledge to planners and policy makers in developing country governments and funding institutions, so that they will perceive less risk in investing in telecommunications. Strategies may include more effort to develop financing packages that will be attractive for smaller, more innovative equipment suppliers and for countries concerned with increasing technology transfer and in-country production or assembly of equipment. They may also include allocating more resources--both multilateral and bilateral--for technical assistance to help countries improve the planning and management of their telecommunications systems, and thereby to increase their net revenues.

Another approach may be to highlight the risk of not investing in Third World telecommunications. To the extent that telecommunications is a complement in the development process and actually contributes to economic growth, low investment may retard economic growth and promote inefficiency. The importance of information as outlined earlier in this paper suggests that underinvestment in telecommunications networks which are the links for information transfer may result in widening of information gaps--both between developed and industrialized countries and within developing countries themselves. The result will be not simply the continuation of current inequities, but the risk of perpetual impoverishment, as those without knowledge about how to improve their condition will continue to lack the means to acquire and share that knowledge.

CONCLUSION

Telecommunications can help to increase the efficiency of economic activities, improve the effectiveness of social services, and extend the social and economic benefits of development more equitably throughout developing societies. Although more research could help to specify the best mix of telecommunications and other development investments, we may conclude at this juncture that there is sufficient understanding and verification of the developmental role of telecommunications to assign a high priority to extending the benefits of telecommunications throughout the developing world. The risks of making this commitment are deterring investment. Yet the risks of not providing universal access to telecommunications may be much greater.

REFERENCES

1. Pierce, William B. and Nicolas Jequier. Telecommunications for Development. Geneva: International Telecommunication Union, 1983.

2. Saunders, Robert J., Jeremy J. Warford, and Bjorn Wellenius. Telecommunications and Economic Development. Baltimore: Johns Hopkins University Press, 1983.

3. Hudson, Heather E., Douglas Goldschmidt, Edwin B. Parker, and Andrew P. Hardy. The Role of Telecommunications in Socio-Economic Development: A Review of the Literature with Guidelines for Further Investigations. Geneva: International Telecommunication Union, 1979.*

4. Saunders et al. Telecommunications and Economic Development. (2)

5. Goldschmidt, Douglas. "Telephone Communications, Collective Supply, and Public Goods: A Case Study of the Alaskan Telephone System." Unpublished Ph.D. dissertation, University of Pennsylvania, 1978. Quoted in Hudson et al. The Role of Telecommunications in Socio-Economic Development. (3)

6. Hudson, Heather E. When Telephones Reach the Village: The Role of Telecommunications in Rural Development. Norwood, NJ: Ablex, 1984.

7. Pelton, Joseph N., Marcel Perras, and Ashok Sinha. INTELSAT: The Global Telecommunications Network. Honolulu: Pacific Telecommunication Conference, 1983.

8. Tyler, Michael et al. "The Impact of Telecommunications on the Performance of a Sample of Business Enterprises in Kenya." Geneva: International Telecommunications Union, 1983.*

9. Gorelik, M.A. and I.B. Efimova. "The Economic Efficiency of Long Distance Telephone Communications." Vėstnik Sviazi, No. 5, 1977. (In Russian). Quoted in Pierce and Jequier, Telecommunications for Development. (1)

10. Hudson, Heather E. "Three Case Studies on the Benefits of Telecommunications in Socio-Economic Development." Geneva: International Telecommunications Union, 1983.*

11. Kaul, S.N. "India's Rural Telephone Network." Geneva: International Telecommunications Union, 1983.*

12. Kamal, A.A. "A Cost-Benefit Analysis of Rural Telephone Service in Egypt." Geneva: International Telecommunications Union, 1983.*

13. Tyler, Michael, R. Morgan and A. Clarke. "Telecommunications and Energy Policy." Geneva: International Telecommunications Union, 1983.*

14. Hardy, Andrew P. "The Role of the Telephone in Economic Development." Telecommunications Policy, Vol. 4, No. 4, December 1980. Also Ph.D. dissertation, Institute for Communication Research, Stanford University, 1980.*

15. Hudson, Heather E., Andrew P. Hardy, and Edwin B. Parker. "Projections of the Impact of Installation of Satellite Earth Stations on National Development." Telecommunications Policy, Vol. 7, No. 4, December 1983.*

16. Hudson, Heather E. When Telephones Reach the Village. (6)

* These studies were sponsored by the International Telecommunications Union (ITU) and the Development Centre of the Organization for Economic Cooperation and Development (OECD). They may be obtained from the ITU.

FINANCING AND OTHER ISSUES IN TELECOMMUNICATION DEVELOPMENT

by

Dr. Edwin B. Parker

Vice President, Equatorial Communications Company

Let's begin by stepping back from the immediate topic of th is conference to the global economic context. The United States and other so-called developed countries are in the early stages of an information revolution, a revolution as profound as the industrial revolution and potentially as socially disruptive because of the more rapid pace of the transition. One symptom of this transformation in America is the unemployment and economic stagnation of the so called "Rust Belt" of historic industrial America. The second major symptom is the high risk and high reward of the booming but erratic growth of "Silicon Valley" and Sun Belt centers of computer, telecommunication and other high technology enterprises. Already the electronics industry is the largest U.S. industry and the growth is continuing. On the average, the economy is booming there, but the layoffs from the shakeout in the personal computer and other high tech industries still disrupt the lives of those directly affected. The more painful transformation of the old industrial sector into more automated, high technology, information-controlled manufacturing will lead to higher industrial productivity, just as the industrial revolution led to the amazing productivity of the highly industrialized agricultural sector of the U.S. economy. But the transition will not be smooth.

The impact of this transformation on developing countries trying to make the transition from pre-industrial agricultural and resource-extraction economies to the industrialized economies will be at least profound and at most devasting. We may be nearing the end of the short period of developed industrial societies taking advantage of lower Third World labor rates by off-shore manufacturing. The alternative of automated production under control of highly educated labor forces in the U.S., Japan and other developed countries may be economically and politically more attractive. The developing countries may not have the luxury of an historic industrial revolution; they may need to leapfrog into the information age in order to achieve even the modest gains in industrial and agricultural productivity that are essential to economic development.

The implementation of telecommunications infrastructure throughout developing countries will be as essential to development as railroads and highways were to developed economies. I won't take time here to review the body of evidence, much of it gathered by Professor Hudson, Professor Hardy and others here at The University of Texas at Austin, all of which quite conclusively demonstrates a very high rate of return on telecommunications investment (1, 2, 3). There is still need for analyses comparing telecommunication investments with other potential uses of funds, and for analyses of optimum complementary investments to achieve higher combined benefits from regional development combining telecommunication with other development investment. For today's purpose we will just assume a highly satisfactory economic return on telecommunication investment.

For purposes of today's discussion, let's also assume the near term availability of appropriate technology, both satellite and terrestrial. Many deliberations on the topic of telecommunications for development make the mistake of treating technology choices as though they were the fundamental policy question. Nothing could be further from the truth. As I have discussed in a recent paper (4), the key policy issue is institutional, not technical. We need policies that lead to initial consensus among third world borrowers, international lenders and telecommunications equipment manufacturers that can allow a viable developing country market for appropriate telecommunications technology to emerge. I have argued the merits of satellite technology for development elsewhere (5) and co-founded a satellite communications company to implement those convictions. So let's assume the timely availability of appropriate technology and skip to the heart of the matter.

The fundamental question is how can developing countries pay for the capital investment in the necessary telecommunications equipment. The moral we draw from our careful research is that in order to achieve economic development, developing countries should invest in telecommunication infrastructure. That advice may be analogous to telling a hungry illiterate subsistence farmer that he would eat better if he were an engineer. It may be true, but quite irrelevant. If there is no way to pay for the education (or the telecommunications investment), the benefits cannot be achieved.

One place to look, of course, is the present cash flow that may be generated by existing telecommunications services (usually urban telephones). Much of the revenue may be needed to pay the principal and interest on the debts incurred when present systems were installed. Net cash available after expenses and debt services could be made available for network expansion. Of course, if that net cash flow is currently being used to subsidize a postal service (as it often is), or for other government purposes, then other funds may need to be found for those other purposes if telecommunications net cash flow is to be reinvested in telecommunications.

Assuming the cash is available, much of it may be earmarked for further expansion of urban telephones. The overall economic benefit to the country may be less than comparable investment in rural services, but the internal cash return to the telecommunications entity may be greater. Besides, the political pressure for more urban telephones may be stronger and there is less political, technical and economic risk in doing more of what they already know how to do. If they don't spend money on expansion of the current urban telephone system, network congestion and breakdowns may result, leading to awkward consequences for the telecommunications managers.

But let's be optimistic; let's assume that internally generated funds are available to invest in expansion of telecommunication services to the more rural or more remote areas of the country previously unserved. Those available funds are almost certainly in the local currency of the country. Most developing countries will be unable to buy appropriate telecommunications technology with local currency. The foreign suppliers will require payment in a currency that can be converted to U.S. dollars, Japanese yen or other national currency of the supplier. So even with cash it is still necessary to convince national economic policy makers to allocate enough foreign exchange. For many countries that are already burdened with foreign debt and a depressed global commodities market, almost all of the foreign exchange generated by exports may be needed merely to pay the interest on previous foreign debt.

This shortage of foreign exchange is at the heart of some difficult economic development policy issues. In principle, if the telecommunications investment can contribute to more export earnings from tourism, agriculture or resource exports, then the necessary foreign currency could be borrowed to buy the necessary telecommunication equipment. Timing is a problem, and there are risks involved, but if lenders and borrowers both believe the loan plus interest can be repaid out of the economic expansion, then there is a chance the funds can be found.

Emerging industrial economies face a more complex choice. In order to stimulate local industry and conserve scarce foreign exchange, foreign imports may be taxed or otherwise discouraged. Local manufacture has much in its favor, including local availability of necessary skills, service and spare parts. If capital equipment is imported when times are good, the shortage of foreign exchange or the currency devaluation resulting from then next recession may render that capital equipment useless if spare parts and service are not locally available.

There may be a fundamental conflict between telecommunications development policy and industrial development policy. The lowest cost and most appropriate telecommunications technology is likely to be based on the high technology of the microprocessor and other silicon or gallium arsenide semiconductor chips. Older technologies have much high costs, higher power consumption, more difficult maintenance and less satisfactory technical performance. Yet the economies of semi-conductor manufacturing depend on the availability of global markets, which make them uneconomic for developing country manufacture on a purely national scale. Consequently, policies that encourage development of local manufacturing may lead to much higher costs, less appropriate technology, and much longer implementation delays while waiting for local manufacturing capability.

The choice is not an easy one. There are merits on both sides of the dilemma. One possible resolution would be based on a comparative return on investment analysis taking into account the economic benefits of both telecommunications development and local industrial development. My intuition is that in cases where a very strong economic benefit case can be made for telecommunications development, then it would be preferable to proceed quickly with the lowest cost most appropriate technology, even if this implies imported technology. The larger economic benefit of telecommunications serving all sectors of a society may exceed the specific benefits of local telecommunications manufacture. In practice, some compromise between imports and local manufacture must be reached. I am merely indicating here that the question does not have easy answers.

With that said, let's return to the central question of where developing countries can obtain the foreign currency for investment in the portion of telecommunications equipment that must be imported. Because there are significant economic benefits to the exporting country, one might assume that the policies of exporting countries would be such as to solve the problem for the importing developing country. From the perspective of a U.S. company that believes it understands the needs and benefits of appropriate telecommunications technology, it would be comforting to think that U.S. policies could help solve the problem. U.S. telecommunications exports would be good for the U.S. balance of trade, create new markets, maintain U.S. employment. There is an opportunity for global economic growth benefiting all parties. Unfortunately, advocates of such policies in the U.S. do not have well financed Political Action Committees or strong established constituencies, compared to those lobbying for restriction of imports of steel, textiles and other exports from developing countries to the U.S. In the present U.S. political climate, it is more likely that government support would be found for the export of older, more expensive, less appropriate technology than for current, lower cost, more appropriate technology. This is because there is excess manufacturing capacity for older technology which no longer has a U.S. domestic market. Because an existing large industry with extensive political power can effectively argue for protection of U.S. jobs, they may be effective. On the other hand, newer technologies, based on the latest microprocessor technologies, may have export barriers put in their way to keep strategic U.S. technology from falling into the hands of communists. Countries other than the U.S. may be able to find ways to help developing countries with appropriate export financing, but they are less likely to have the most appropriate, lowest cost technology available.

A variety of potential loan sources may be available for rural telecommunications development, provided a number of troublesome policy and management issues can be resolved (see 6). For either commercial banks or international development banks to make the necessary funds available, however, they must satisfy themselves that the risks are not too great. No matter how good the theoretical rate of return on the economic investment, the practical concern is that poorly executed projects will not achieve the theoretical benefits. In the real world there are many places to go wrong, many risks and many opportunities for failure.

Which institutions in the developing countries have the necessary human managerial resources to avoid mistakes in procurement of new technology, in familiar installation programs, in the complicated logistics of managing rural programs, and in the myriad operational details of operations, billing, sparing, maintaining, and training. One of the reasons why rural telecommunications is economically essential is that without telecommunications, any rural development program or project, whether government or commercial, is extremely difficult to manage. A rural telecommunications installation program will suffer the same problems until the telecommunications equipment is effectively in operation. Many logistics problems may go unresolved until somehow success is achieved. The installation and maintenance of rural telecommunications will not be easy, because, by definition, the reliable telecommunications network won't yet be in place to help overcome the inevitable logistics problems.

So who will take the risks of trying to put it all together? Who has the incentive to make it all work? At a time of rapid technological change, which will be the case in telecommunications for the next 25 years, who will decide when the newer lower cost technologies are proven enough to be risked? Telecommunications no longer means just plain old analog voice telephone circuits, but now includes telex, telegraphy, electronic mail and other data transmission, as well as broadcast media and teleconferencing services. Who is going to make the necessary judgments concerning which of many services shall be provided? Who can take the risks of cost overruns, schedule delays or unpredictable political changes?

Present telecommunications institutions established as government monopolies may, with reason, be too risk averse to take on all the challenges, especially if they have vested interests in older technologies and older procedures which they understand and know how to manage. They may not be able to adapt quickly enough to the new challenges and opportunities. Perhaps they would be right to duck some of the inevitable new problems.

In the United States, where the economic resources are available and the new telecommunications technologies are being created, it was decided that leaving telecommunications development to the prior monopoly was too risky. Monopoly worked well during times of stable or slowly changing technology. But now the choices of both telecommunications technologies and services are many. The opportunity for dramatic productivity gains and consequent national economic growth is there, but risks come with it. The U.S. government has determined that the inevitable turmoil created by a major institutional change in telecommunications is a small price to pay to achieve the economic growth expected from new technology and service development stimulated by competition.

I am not arguing that U.S. style competition is appropriate for poor developing country telecommunication sectors. I do, however, disagree with the conventional wisdom of assuming that rural telecommunications technology should be similar to urban technology, and financed by cross-subsidy from monopoly revenues elsewhere. That approach served the U.S. well, and led to the present near-universal service in the U.S. But developing countries can not afford those kinds of costs and can't generate that much cross-subsidy revenue. Therefore, we need to encourage an institutional structure that can stimulate innovative, lower-cost appropriate rural telecommunications technologies without imposing excessive risks on the national telecommunications monopoly.

Since the need and the demand are so large relative to the potentially available supply, protecting a monopoly by preventing other government agencies or the private sector from establishing essential services may be too short sighted. Perhaps instead, any public or private entity attempting to bring needed service to areas that the monopoly telecommunications entity cannot serve, should be encouraged instead of prohibited. Policies encouraging local public access to and interconnection of private networks to the public network may be appropriate to achieve legitimate telecommunications policy goals. In that way entities with the strongest needs and incentives could be encouraged instead of prohibited from risking telecommunications development. The external economic return on telecommunications investment is most likely to be greatest when it is implemented as an integral part of whatever economic development activity is going on in the previously unserved region.

In that way, developing countries can have some experience with rural telecommunications projects managed by institutions with the greatest incentive to be successful. As a result of a variety of such experiences, the monopoly telecommunications entity may gain, at least vicariously, the local experience necessary to begin extending service to the rest of the country.

Once there are a few successful projects as models to follow, and experience has been gained with alternate technologies designed for thin-route rural telecommunications, then the risks will be less for larger scale programs. With lower risks, it should then be easier to achieve the consensus among developing country borrowers international lenders and telecommunications manufacturers necessary to make large scale funding readily available.

REFERENCES

1. Hardy, Andrew P. "The Role of the Telephone in Economic Development. Telecommunications Policy, Vol. 4, No. 4, December 1980..

2. Hudson, Heather E. "Three Case Studies on the Benefits of Telecommunications in Socio-Economic Development". Geneva: International Telecommunication Union, 1981.

3. Hudson, Heather E., Andrew P. Hardy, and Edwin B. Parker. "Projections on the Installation of Satellite Earth Stations on National Development". Telecommunications Policy, Vol. 7, No. 4, December 1983.

4. Parker, E.B. "Appropriate Telecommunications for Economic Development". Telecommunications Policy, Vol. 8, No.3, September 1984.

5. Parker, E.B. "Communication Satellites for Rural Service". Telecommunications Policy, Vol. 5, No. 1, March 1981.

6. Goldschmidt, Douglas. "Financing Telecommunications for Rural Development", Telecommunications Policy, Vol. 8, no. 3, September 1984.

BRAZILIAN SATELLITE COMMUNICATIONS PROGRAM

by

Joao Carlos Fagundes Albernaz, Ph.D.

Secretary of Informatics

Ministry of Communications

Brazil

INTRODUCTION

The main objective of the process of modernizing and expanding telecommunications services to Brazil is to provide the country with a telecommunications infrastructure commensurate with its stage of development, incorporating the technological advances achieved by this sector in other parts of the world. This incorporation must always take into account the characteristics of Brazil, in such a way that the modernization and expansion of the services will ensure the well-being of the population and the economic growth of the country.

Brazil, with an area of approximately 8.5 million square kilometers, is the largest country in South America and has 128 million inhabitants, the sixth largest population in the world. This population is concentrated in the south central area, where the largest cities and most of the industries are located.

The Brazilian Amazon Region, which encompasses about 59% of the territory, is covered by tropical forests and some areas are accessible only by plane or riverboat.

The Brazilian Amazon Region extends over 5 million square kilometers and has a population of 8.5 million people, corresponding to a demographic density of 1.7 inhabitants per square kilometer. Most cities in the region are in reality small villages with less than five thousand people each, in contrast to the cities in the South-East region of the country where a great part of the country's population is concentrated.

Since 1972, microwave line-of-sight terrestrial systems interconnect all state capitals and the main cities in the South-East, North-East and Central regions of the country. Until 1974, communications with the state capitals in the Amazon Basin and other cities in that area was effected via troposcatter radio relay systems. Geographical dispersion of the population and geographical peculiarities precluded the use of line-of-sight terrestrial systems for provision of telephone and TV service.

It was apparent that by taking advantage of the characteristics inherent in a satellite network, it would become possible to link up cities and villages in the Amazon Region with the rest of the country.

THE BRAZILIAN SATELLITE COMMUNICATIONS SYSTEM

Communication satellites are modern and efficient means of telecommunications that may meet telecommunications needs of some developing countries. They usually represent the best cost effective alternative for the provision of telecommunications services to geographically dispersed and isolated areas.

Countries such as Brazil, India and China which have in common extensive territorial areas, and less developed cities in remote regions with difficult accessibility, are likely candidates to implement communication satellite systems aiming to:

* support national programs in areas related to education and health assistance;
* extend the broadcasting coverage area of the country, especially TV broadcasting, offering a choice of programs (more than one channel for each city);
* carry public telecommunications services to remote areas and supplement terrestrial links.

In the case of Brazil, besides the potential benefits that may accrue to the sectors of health, education and agriculture, the use of satellite communications has been for some time an important factor for the integration of the Amazon Region with the rest of the country through the use of telecommunications and also for relaying TV and radio program to inland cities.

The benefits offered by this modern means of communications and the need to extend telecommunications facilities to regions where geographical conditions would become difficult or even hinder the use of more conventional means led the Brazilian Government to the decision to establish a domestic satellite communications system.

INITIAL PHASE

In 1965, Brazil became a member of INTELSAT (The International Telecommunications Satellite Organization) and since 1969, Brazil has utilized INTELSAT satellites for international communications. Currently Brazil is the fifth largest user of the INTELSAT system.

In 1974, EMBRATEL inaugurated the Brazilian domestic satellite communications system using transponders leased from INTELSAT satellites. Studies undertaken at that time showed the feasibility of using a Brazilian owned satellite for domestic communications. A project was designed and EMBRATEL was authorized to take contract bids from overseas firms to implement it. For economic reasons, however, the bidding process had to be suspended.

Since then, the use of satellite communications has been intensified in Brazil. Today 21 earth stations are in operation in the Amazon Region and one on the island of Fernando de Noronha. EMBRATEL'S earth stations are equipped to receive and transmit telephone signals and receive TV signals. The stations in Rio de Janeiro, Sao Paulo and Manaus are also equipped to transmit TV signals.

It is worth mentioning the increase in telecommunications traffic in three towns in the Amazon Region after the installation of local earth stations, in 1981, when they became operational.

Table 1 shows that the telecommunications traffic has increased up to 20 times, demonstrating the expansion potential of the BRASILSAT system. From this table we could also infer the economic and social benefits that have resulted from the use of satellite communications to the local population.

In 1982, the Ministry of Communications opened up the use of earth stations for TV reception only (TVRO), and in recent months Brazilian banks have started using SBTS - Brazilian Satellite Communications System - for high speed data transmission. Today more than 400 earth stations are dispersed throughout the country, with this figure continuing to increase.

Studies conducted by EMBRATEL reveal that in the Amazon Region, about 1,200 villages and towns are potential candidates for some form of satellite communications. The market estimate for TVRO stations in Brazil is 3,000 units.

Growth in the use of satellite communications in Brazil may be observed in Tables 2 and 3. The number of transponders leased from INTELSAT satellites has increased from 1 in 1974 to 8 in 1984. Two are global transponders and six are hemispheric.

DECISION TO OWN A SATELLITE SYSTEM

As the traffic volume increased, it became necessary to lease more transponders from INTELSAT, thereby increasing the rental costs. Because of this increasing cost, in 1979, the Ministry of Communications re-examined earlier studies regarding the placing in orbit of two Brazilian owned communications satellites. These satellites, designed to provide complete coverage of Brazilian territory, would permit the use of smaller and cheaper earth stations as compared with the ones used in conjunction with INTELSAT satellites.

The economic aspects of these studies also showed that the cost of buying and launching two satellites would compare favorably with the estimated cost of leasing transponders in INTELSAT satellites over a ten year period. Furthermore, today's use of INTELSAT transponders contemplates telephone signals and TV transmission only.

Recent market surveys conducted by EMBRATEL have shown that the more promising market for satellite communications in the country in the years ahead are TV reception, rural telephony and private data communication networks. These applications demand in most cases low cost earth stations equipped with low power transmitters, SCPC equipment and small diameter antennas, which can be locally designed and manufactured.

NEW USES FOR SATELLITE COMMUNICATIONS

A Brazilian owned satellite would open up new areas of application:

a) Public Telecommunications

Inland localities not served by telecommunications could be interconnected to the NTS (National Telecommunications System), as strategic or social reasons would require. The possibility of extending satellite use to rural areas and to newly-developing areas defines new potential users of the new means of telecommunications in the short term. Due to the flexibility of the satellite communications system, it is well-suited to high speed data communication. In a more advanced phase, the satellite communications system will enable the expansion of the NTS itself, in alternate or complementary routes.

b) Transmission of Television Signals

In Brazil the use of direct broadcasting satellites has been under examination for some time. It was concluded, at least for the first series of Brazilian satellites, that this application would render the system unfeasible, due to the necessity of high transmission power. For this reason, insofar as TV reception was concerned in the present stage of development of space technology, the Brazilian Government opted for community reception only, thus necessitating the use of reception equipment in earth stations and transmission of the received signal to TV stations or to re-transmission networks. Certainly, the existence of a domestic satellite communications system leads to the formation of large television networks, thus freeing up channels in the terrestrial microwave links for other more convenient uses.

c) Special Applications

The domestic satellite communications system, besides servicing new users, will also render other services possible, including:

* extension of the national telecommunications system to areas of difficult access by conventional means, such as oil exploration and drilling platforms, agro-industrial sites and mining centers;
* support for social services - tele-education, tele-medicine;
* servicing of temporary communications necessities in work sites, such as dam constructions and construction firm work sites;
* establishment of private networks for large users, such as banks and public and private entities of regional and national scope;
* inland transmission and distribution of radio and television programs;
* remote and simultaneous printing of newspapers and periodicals;
* teleconferencing;
* data collection, remote sensing and others.

ECONOMIC FACTORS

The studies that preceded that action of the Ministry of Communications, leading to the establishment of a Brazilian communications system via satellite, also included an analysis of economic factors. These studies considered the following uses for the system: telephone and telex services, transmission of television signals and data transmission to places of difficult access. These uses account for the bulk of Brazil's telecommunications traffic, representing more than 95% of total revenues.

Also considered were the expenditures in foreign currency, with the implementation of the space segment, as compared with the expenses with the rent of transponders from INTELSAT for domestic use. Finally, the cost of the domestic satellite program was compared with the cost of other programs carried by the TELEBRAS group.

The analysis led to the following conclusions:

* the resulting savings on expenditures with the leasing of transponders would amortize the cost of the Brazilian owned space segment;
* the annual amortization of the investment in the satellite would not represent a significant share in the capital expenses budget of TELEBRAS.

SATELLITE OWNERSHIP OR TRANSPONDER LEASING?

Comparison between leasing transponders of INTELSAT satellites and owning the space segment favors the latter alternative.

a) Leasing Transponders from INTELSAT

Advantages:

- makes use of only the necessary capacity;
- no launching expenses;
- future possibility of purchase of own satellite.

Disadvantages:

- earth segment must be adjusted to INTELSAT specifications;
- higher costs of earth stations;
- use of transponders limited to INTELSAT operating procedures;
- non-optimized geographic coverage.

b) Own Space Segment

Advantages:

- space segment adjusted to the country characteristics and needs;
- lower costs of earth stations;
- transponder use not limited to INTELSAT operating procedures.

Disadvantages:

- low utilization of the space segment in the first years of operation.

BRAZILIAN COMMUNICATIONS SATELLITE - BRASILSAT

In February 1981, the Ministry of Communications was authorized to carry out the project.

An international bid was the basis for the selection of suppliers for the satellites and the launch operation. Some specific requirements were to be met by the suppliers:

* a counterpart of additional export of Brazilian manufactured goods equivalent in value to that of the products or services rendered;
* financing conditions;
* technology transfer;
* previous experience in the area.

In July 1982, EMBRATEL signed contracts with a Canadian Company, SPAR, which led a consortium of companies with the participation of Hughes to supply the satellites and with a French enterprise, ARIANESPACE, to provide the launch vehicles and services. ARIANESPACE was the only bidder capable of meeting the requirements set out in the bidding timetable for the launching of the first satellite.

The main points included in the contracts between EMBRATEL and the suppliers are:

a) SPAR/EMBRATEL Contract:

OBJECT: - 2 satellites

- Telemetry, Telecommand and Control Earth Station - TCC Station

COST: - US$ 122 million without orbit incentives

- US$ 131 million with orbit incentives

FINANCING:

- EDC/CANADA - US$ 80 million
- SCOTIA BANK - US$ 15 million
- EXIMBANK - US$ 27 million

OTHER OBLIGATIONS:

- additional financing from the Royal Bank of Canada: US$ 15 million:

The main points included in the contracts between EMBRATEL and the suppliers are:

- export counterpart in the value of US$ 200 million within a 4 year period;
- program of transfer of technology to the TELEBRAS Center of Research and Development (CPqD) and to the Brazilian Space Research Institute (INPE).

b) ARIANESPACE/EMBRATEL Contract:

OBJECT: - 2 launch vehicles and associated services

COST: - US$ 58 million

FINANCING:

- US$ 40 million from Credit Lyonnais/Paribas - France

OTHER OBLIGATIONS:

- additional financing: US$ 46 million from Credit Lyonnais/Paribas;
- program of technology transfer to the Brazilian Space Research Institute (INPE) and to the Brazilian Aerospace Institute (IAE).

General characteristics of the Brazilian satellite are shown in Table 4. The TTC earth station is located in Guaratiba in the State of Rio de Janeiro. The satellites will be launched from the Kourou Space Center (French Guiana) in February and August 1985.

The locations of the satellites in the geostationary orbit will be 65 ° W (BRASILSAT I) and 70 ° W (BRASILSAT II). Beginning of operation of BRASILSAT I is scheduled for March 1985.

TECHNOLOGY TRANSFER

The introduction of a domestic satellite communications system has followed a pattern that has been observed in the expansion of telecommunications networks in Brazil.

In the first phase, "turn-key" systems had to be imported. However, since then, Brazilian technicians have been able to learn how to operate and maintain these systems.

In the second phase, local industries started to produce some of the equipment and even to introduce changes in the original design of the equipment to allow the use of Brazilian made components in the manufacturing process and to make it more compatible with local conditions.

In the final phase, the main efforts are directed towards research and development, and creating the essential conditions for the establishment of an adequate level of technological self-sufficiency in the country.

It is my opinion that such a pattern deserves the attention of other developing nations interested in the process of technology transfer.

In Brazil, where the state-owned telecommunications service companies are the major equipment buyers, the Ministry of Communications has a major influence on Brazil's industrial policy for telecommunications. This influence has been used to stimulate the development of a national telecommunications equipment industry, to decrease dependence on foreign suppliers, to permit an adequate level of market competition and to ensure a significant level of decision-making autonomy regarding matters related to industrial and technological development in the telecommunications industry.

Developing and consolidating a national industry and local technological capability serves to strengthen the corresponding industrial structure, resulting in a greater degree of national autonomy. Of course these objectives are to be harmonized with operational objectives aiming to provide the best possible service to the public.

The establishment of a domestic satellite communications system in Brazil has followed the above mentioned pattern.

The first earth station used for international communications was bought and installed on a "turn-key" basis in 1969, but has always been operated and maintained by Brazilian technicians. Although the first few earth stations for use in the domestic satellite system were supplied by foreign manufacturers, their system specification and installation were the work of EMBRATEL's engineers. Furthermore, with the decision to start the operation of a domestic communications satellite system, a joint development program was established with the participation of TELEBRAS/CPqD and local industries aiming at developing and manufacturing complete earth stations in the country.

This program, which was initiated in 1974, has led to some major achievements, as shown in Table 5. More improvements will follow as the volume of production increases. It must be recognized that some technological barriers will remain with respect to some specific products. Nevertheless, some degree of dependence will always remain, for economic reasons if for no other.

While TELEBRAS Research and Development Center for local industries successfully developed most of the parts and equipment necessary to the earth segment of the Brazilian Satellite System, this was not the case insofar as the launch vehicle and the satellie were concerned. Nevertheless, programs are being conducted in the country by the Space Research Institute (INPE/CNPq) and by the Aerospace Research Institute, in Sao Jose dos Campos, Sao Paulo, aiming at the launching and placing into orbit of a Brazilian designed and manufactured remote sensing satellite. The launch vehicle will also be locally designed and manufactured.

For these reasons, specific requirements were established in the bidding regarding the transfer of technology related to the satellite, launch vehicle and associated services. This will accelerate the aforementioned programs.

CONCLUSION

The establishment of a domestic satellite communications system in Brazil--SBTS--began in 1974 with the installation of two earth stations in the cities of Manaus and Cuiaba and using transponders leased from INTELSAT.

This system expanded and today 8 transponders are leased from INTELSAT to handle the traffic; 24 EMBRATEL earth stations operate in conjunction with the domestic communications satellite system; two Brazilian TV networks relay their signals through the SBTS for local retransmission in 29 cities; their programs are also received by another 400 small earth stations dispersed throughout the country. New services, notably data communications, are finding the satellite an economically and technically preferable alternative.

The evidence of the importance and of the advantages associated with the utilization of communications satellites in a country with the geographic dimensions and characteristics of Brazil led the Government to the decision of owning its own satellites - BRASILSAT.

Undoubtedly, the beginning of operations of BRASILSAT will coincide with a period of remarkable growth in the use of satellite communications in the country, and consequently, the earth segment will have to be expanded in order to meet the increased telecommunications needs.

Taking this fact into consideration, the Ministry of Communications initiated a program in 1975 designed to stimulate the development and manufacture, by local industries, of the necessary equipment. Because of this program, it will be possible for most of the equipment for the earth stations in the expansion of the earth segment of the SBTS to be provided by local manufacturers utilizing locally developed or adapted technology.

In an economy where the scarcity of foreign currency is a factor, this policy has proved effective inasmuch as it will allow for the expansion of the earth segment without relying on high levels of imports.

A final question remains: what of the future for communications satellites in Brazil?

BRASILSAT will be a most effective tool in the Government programs of popularization and interiorization of telecommunications services. At the present time, the 10.5 million telephones in service in Brazil are largely concentrated in the main cities, and not all segments of the society can afford to own a telephone. BRASILSAT can make rural telephony and other programs such as telemedicine and tele-education a reality for communities in the Brazilian interior.

It is my personal view that in Brazil, local institutions have made some progress in the area of instructional TV, and local industries are now capable of producing low cost earth stations given the proper scale of production. The next Government could give serious consideration to this proposal.

I hope that in a conference like this, within a two year timeframe, much more will be said related to the use of BRASILSAT in connection with tele-education, rural telephony and data communications.

Table 1: Traffic Growth after the Installation of Local Earth Station (Typical cases - 1981)

CITIES	TRAFFIC Thousands of minutes per month		INCREASE
	Before	After	
TABATINGA	8.6	172	1:20
TEFE	10.8	114	1:10
COARI	10	85	1:8

Table 2: Growth in the Use of Satellite Communications in Brazil (number of voice channels)

	1979	1981	1983
FDM	816	1,692	1,974
SCPC	-	308	632
TOTAL	816	2,000	2,606

Table 3: Number of Earth Stations Owned by EMBRATEL (Domestic System)

		1979	1981	1983
TELEPHONY	FDM	4	3	4
	SCPC	-	9	12
	FDM & SCPC	-	4	4
TVRO		2	1	1
TOTAL		6	17	21

Table 4: BRASILSAT - General Characteristics

STABILIZATION	spin-stablized
DIMENSIONS	overall height: 2.95m (launch)
	7.09m (in orbit)
	diameter: 2.16m
WEIGHT	1,140 kg (launch)
	670 kg (in orbit/beginning of life)
LAUNCH VEHICLE	ARIANE III
POWER	982 W (beginning of life)
	799 W (end of life)
LIFETIME	8 to 10 years
OPERATING FREQUENCY	6/4 GHz
NUMBER OF TRANSPONDERS	24 (36 MHz bandwidth)
EIRP	34.0 dBW

Table 5: Development and Manufacture of Earth Station Equipment in Brazil

EQUIPMENT	TECHNOLOGY	PERCENTAGE OF IMPORTED PARTS AND COMPONENTS
ANTENNA 3m	Locally Developed	-
4/5m	"	-
6m	"	-
10m	"	5%
LNA 90 K	Locally Developed	50%
DOWN CONVERTER	Locally Developed	50%
SCPC	Locally Developed	20%
FDM	Foreign	15%
MESSAGE EXCITER-FDM	Locally Developed	30%
MESSAGE RECEIVER-FDM	Locally Developed	10%
HPA 100 W	Locally Developed	50%
TV RECEIVER	Locally Developed	10%
ECHO SUPPRESSOR	Foreign	15%

ISSUES IN TELECOMMUNICATIONS TECHNOLOGY TRANSFER: A WORLD BANK PERSPECTIVE

by

Richard Stern

Chief, Telecommunications, Electronics and New Technology and New Technology Development Division

The World Bank

What I'd like to do is to give you my perspective of the less developed country environment and the framework within which the specific problems and issues being addressed by this conference must be considered. I would adhere to the view expressed earlier by Edwin Parker that the issues we face in less developed countries in telecommunications largely fall into the institutional, financial, and manpower categories. Paradoxically, technology issues, the subject matter of this conference, may be of secondary importance. What's new on the technological side is that we now have many and exciting options which were not available to us in the past. However, it is important to consider the country and institutional environment in which these choices will be made.

Just for a minute, let me remind you a little about the environment we are discussing. The Third World has about 7% of the world's 600 million phones. The average telephone density is about 2.8 per 100 population. That compares to 84 per 100 population in the U.S. Even this low average figure overstates the paucity of the networks in some countries. For example, the average density in many African countries is around 0.1 telephones per 100 population. In India and Pakistan, it's about 0.4 per 100 telephones. By contrast, in Argentina, it's 10.3 and in Brazil somewhat higher than that. Even these figures hide the dramatic concentration in the cities. Ethiopia is a country that is generally regarded as a pioneer in the development of telephony in Africa, yet the rural density is less than 0.08 per 100 population. India has given high priority to rural telephony yet the density is only 0.2 per 100; in significant parts of the Philippines, rural telephone service is nonexistent. The LDCs are thus characterized by an extreme paucity, or absence, of telephones where the populations have to resort to costly substitutes to make up for the lack of this essential infrastructure.

Most of the entities that I am familiar with have also been characterized by slow rates of growth in their services, often not much more than growth rates in their overall economies. As Heather Hudson mentioned, the development of telephone services has typically been given low priority by governments, telephony being regarded as a consequence of economic development rather than a major cause. Telecommunications also suffered, as was mentioned earlier, in the competition for scarce financial resources because of its high foreign exchange content. The foreign cost of telecommunications investment projects is many LDCs is of the order of 70%, 80% or even 90% of total costs. Comparable ratios for the agricultural sector typically run around 25%-40% of total investments. This is the environment in which the promoters of telecommmunications investments in less developed countries have to bid for resources.

Reflecting the above, telephone systems in LDCs are also typically small. In Rwanda and Burundi, we're talking about total system sizes of the order of 6,000 phones; in Ethiopia, 60,000 phones; in Bangladesh, somewhere between 85,000 and 90,000. Indonesia, the fifth largest country in the world, with 150 million people, has about 240,000 phones. To put this in perspective, if I read my guide to Austin right, that would leave Indonesia and greater Austin with roughly the same size phone system!

These small systems mean that fixed costs cannot be spread over a large number of phones. The implied paucity of the networks also means that entities must be very risk adverse in considering new investments and making technological choices. If a new state of the art system goes on the blink in the developed world, there are always alternative systems available to route telephone calls; in the typical LDC network there will be none. I think this fact has special meaning when considering the application of satellite systems -- the subject of this conference. The critical questions facing these countries in making technological choices may therefore be very different from those facing a developed country.

What else is characteristic of LDCs? They're operating in an environment of substantial excess demand, a demand that won't be met for the foreseeable future. This is a major contrast to the developed countries. Here is a major contrast to the developed countries. Here in the U.S. environment and in Europe, telephone companies are trying to sell new systems and services to markets where POTS -- plain old telephone service -- has already been provided. That's a very different situation from planning telecommunications in situations where prospects of just meeting the demand for basic service in the near or medium term are somewhere between minimal to nonexistent. Incidentally, I know this is the subject of intense discussion, and there are obviously some exceptions, but my own feeling is that the highest rates of return to investments in LDCs lie in basic voice communication for a long time to come. Again, this is a very different environment form that faced by most of the people present at this conference.

These countries are also characterized by very poor

quality of service. Equipment is often old and long depreciated. I was in a telephone exchange in Arusha in Tanzania four weeks ago. I think I counted seven different types of terminal equipment, the first items having been installed in the early 1950s, the last pieces of equipment being put in only a year or two ago. Somehow the technicians keep it working, but with daunting and mounting spare part and logistic problems. Antiquated exchanges and decaying cable networks result in systems plagued by faults and subscriber frustration.

In the organization side we typically encounter telephone companies which are extremely weak. They're often not much more than a government department with very little autonomy and reflect many of the problems typically associated with these bureaucracies. They also reflect the pressures to create employment in countries where such opportunities are limited. It's not unusual to have telephone entities in which staff substantially exceeds 100 per 1,000 lines. The equivalent figure for this country is less than seven. But paradoxically these companies are often short of critical skills -- both to plan and manage the new technologies as well as to accomplish some of the more routine tasks. I'll come back to that later.

Despite the fact that LDC telephone companies are typically monopolies operating in an environment of excess demand, it's not uncommon to find entities which are a net drain rather than a net contributor to government resources. This phenomenon reflects poor and unbalanced plant; it reflects overstaffing; it reflects operational inefficiencies; and it reflects inadequate tariff levels. It's not unusual for customers' bills to be sent out after considerable delay. In Ghana, for example, customer billing was the subject of substantial delays until quite recently because of a lack of the right paper on which to print the billing statements! Also, in many LDCs, tariff structures, as opposed to tariff levels, are woefully inappropriate. Your local U.S. telephone company can tell you exactly what happens to system use and revenues if they reduce or raise calling charges at given times of the day. They've done the theory, and they've tested it in practice; they can thus optimize system use and maximize revenues. Very little work on these issues has taken place in less developed countries, despite the fact that optimization of the use of existing networks in LDCs is critically important, considering the resource and foreign exchange constraints facing the LDCs that I mentioned earlier.

While this is a bleak environment, I do see change on the horizon. I think a number of countries are now recognizing past neglect. Whether this movement is a spin-off of the information revolution in the developed world I don't know, but one does now see an increasing realization of the need to redress the situation and give higher priority to telecommunications investments in the Third World. The view that telecommunications is a luxury good is dissipating. Incidentally, I'm not sure that in the past the telecommunications community has done a very good job in selling itself and making the case for telecommunications as an essential infrastructure. It is important that we do all become advocates for our case. A new priority for telecommunications can be seen in Pakistan, in Indonesia, in India, in Kenya and in Ethiopia. These are just a few examples; in recent years these countries have made major policy decisions to give greater priority to the sector.

Once the basic decision to expand is made, what's a realistic growth rate? I'm not sure. I would say somewhere between 10% and 15% a year may be a reasonable target. I think much more than that, while fully justified, is not realistic for most countries. It won't be financed; the resources are just not available. Let's assume that the average cost for telecommunications investments in less developed countries amounts to somewhere around $2,000 for each direct exchange line added to the system. We can argue about this estimate but I think it's a fair number. This number is eight to ten times the average per capita income of many LDCs. Even for a middle income country like Indonesia, this investment cost is something like three times its per capita income. To argue for more resources in such an environment to financed growth rate in telecommunications investment beyond 10% to 15% is therefore probably not realistic.

This scarcity also requires that we chose our investment priorities carefully. For example, while Dick Butler was giving his telephonic address this morning, I was trying to work out a few numbers. If I got my figures right, a 10% to 15% annual expansion program for telecommunications in Sub-Saharan Africa would translate into an investment of our $300 million a year. If you compare this figure to the $800 million to $1 billion figure that he mentioned to put the proposed African satellite into orbit and to construct the related earth stations -- let alone the associated terrestrial network -- our investment priorities may, for the near and medium term anyway, come out somewhat differently. Whether we work to put all our eggs into one technological basket and to take the associated risks -- a point I alluded to earlier -- also becomes a vital question in this financially constrained situation.

Related to these issues is the importance of training and developing the manpower capacity within LDCs to make these choices and trade-offs. The World Bank, USAID, other bilateral programs and the ITU in particular which has been a notable pioneer in this area, recognize and give priority to these issues but I think it behoves all of us to give even more attention to the development of these manpower capabilities. Related to this there's also an equally important need to develop new organizational frameworks and structures for LDC telephone operating entities. You know, if you are expanding at 15% a year, that means you double the size of your system every five years. In turn that requires a fundamental change in the associated institutional structure. The challenge facing us is even more daunting when you consider that we are seeking these far-reaching institutional changes in societies which possess, as one of their principal characteristics, a lack of such institutional dynamism. Even when some countries have managed to meet these challenges, some parts of the systems get left behind. For example, as I understand it, Brazil's urban network has expanded extremely rapidly, and the associated organizational institutional changes that have been enacted to support this process have been quite effective. However, an organizational framework designed to support and finance the rapid expansion of a rural system and cater to its special needs still has to be identified. The importance of developing appropriate organizational, procedural, managerial and associated training systems for LDC telephone entities cannot, in my view, be overemphasized.

Can expansion programs of 10% and 15% per year be financed? I think the answer is yes. This issue is probably best examined from the viewpoint of local and foreign cost requirements. We in The World Bank feel quite strongly that the local resources required for telecommunications expansion programs of this order of magnitude can and should come from self-generated funds. Actually, properly run telephone entities should be able to contribute substantial financial resources to governments to finance development in other sectors as well as meeting their own local financing requirements.

Generating the required foreign exchange is a problem. Given the inherent profitability of properly planned telecommunications investments, attracting private foreign capital in the form of investment and/or loans clearly has potential for several LDCs. However, in many of the countries we are talking about, foreign investors will be reluctant to take on the country as opposed to the project risk. These countries are short of foreign exchange, and have limited creditworthiness. Investors or creditors may thus be unable to convert their earnings from telecommunications investments into foreign exchange and so repatriate their profits. So however high the returns to telecommunications investments, country risk becomes critical. Various insurance schemes designed to protect investors against some of these risks have been suggested and should be pursued further; however, the fundamental problem of country creditworthiness remains.

Some countries have dealt with this problem by promoting the development of a domestic telecommunications manufacturing industry. For example, Brazil, India and Korea have all become major producers. In fact, if I recall correctly, local value added in the Brazilian telecommunications manufacturing industry is over 90%. Put in another way, Brazil now only has to secure foreign exchange financing for 10% of its investment program in the telecommunications sector, i.e., foreign financing is no longer a significant problem. This approach can, however, also result in significant costs. I think a lot of countries have made terrible mistakes by jumping too quickly and overly protecting their domestic manufacturers from competition. A significant portion of LDC telecommunications manufacture is high-cost and low-quality. Often they have tried to become too independent too rapidly, have missed some of the significant cost saving benefits of the rapidly changing and new technologies, and have been trapped with some of the older, more expensive and now obsolete systems. However, for the larger developing countries there seems to be no a priori reason why, with proper planning, that they should not be able to develop efficient and cost effective domestic manufacturing industries capable of meeting a large portion of their telecommunications equipment requirements. While I'm not sure that I would encourage many LDCs to go in for large-scale switching manufacture, a large number of countries would appear to be able to efficiently produce such items as PBXs and some transmission and terminal equipment.

We still have to get down to how to finance the foreign exchange requirements for the majority of countries who cannot, or should not, develop a domestic manufacturing industry. If we achieve a growth rate of 10% to 15% a year in LDC telephony, we are by my quick reckoning talking abut a global requirement of the order of $6-8 billion a year. This is clearly a challenge. Telecommunications entities will have to continue to rely on a mix of such traditional sources as the private investment, private suppliers' credits, commercial banks, multilateral banks, bilateral aid and export credit agencies. New techniques to attract the additional funds also need to be developed. It seems to me the great advantage that we have in the telecommunications community is the obvious interest of the developed world in developing LDC markets for their products. Furthermore, there is significant overcapacity in the manufacturing industry in several countries. In this environment a number of developing countries have been able to successfully organize international bidding for both price and financing. We feel in the World Bank that properly organized competitive procurement is critically important. Our own experience suggests that international bidding in telecommunications results in unit costs which are typically 25% to 30% below bilaterally negotiated prices. Recent experience with the above bidding procedure in Guatemala and Uruguay has also resulted in competitive financing offers for much of the equipment to be supplied. Given the active interest of governments to promote their telecommunications exports, this is clearly an area where LDCs could have some success in developing novel and attractive financing schemes.

However, given the problems of country risk I mentioned earlier, there are clearly limits to the total amount of foreign currency any LDC can prudently borrow on commercial terms for whatever purpose. The level of investment in telecommunications in the poorest and the least creditworthy countries will ultimately depend on the priority their governments attach to telecommunications in allocating their scarce foreign exchange resources and the willingness of the developed world to provide aid or grants for telecommunications: by definition, aid does not become a burden on the recipient's debt service capacity.

To conclude, I hope I have not presented too bleak a picture; that was certainly not my intention. Overall, I feel there are grounds to be optimistic. The priority many

LDCs attach to telecommunications investments is clearly changing for the better. Moreover, the new technologies in the sector, the declining unit costs and easier system maintenance with which they are typically associated, present the developing countries with new and exciting opportunities. However, the environment within which we are operating demands that we proceed with a good dose of realism and care. A rapid build-up in overall levels of investment in telecommunications is unlikely to yield the expected benefits without concomitant attention being paid to institutional, organizational and operational reform. The development of the necessary planning and administrative systems in turn requires massive and well designed training programs. Lastly, the introduction of proper financial accounting systems including appropriate tariff policies is an essential prerequisite to success. Experience in several LDCs has shown that while these problems cannot be resolved overnight, that steady and impressive progress is indeed possible. The resolution of these issues is important to ensure that optimal investment decisions are made and that technology choices are properly considered; particularly with respect to the inability of those LDCs' entities to take undue risk. Lastly, given the appalling scarcity of resources in most of the countries of which we are speaking, they just cannot afford to bear the heavy financial and economic burdens associated with badly planned and ill used investments in telecommunications. Our continuing attention to the resolution of these broader issues is therefore essential.

DISCUSSION

MR. SRIRANGAN'S COMMENTS ON MR. STERN'S PRESENTATION:

Much of the current policy analysis concerning rural telecommunications has not focussed upon examining the telephone density of a country, but rather upon the spatial distribution of telephones. In looking at this question of access, India has set an objective of placing a telephone within 5 kilometers of each person. In addition, we have attempted to contrast urban and rural telephone densities in an effort to study the problem of equitable access.

Mr. Stern's goal of 15% annual growth in telecommunications investment for the developing world is certainly realistic and it is also viable in terms of flowback, or return. One must note the tariff approach in this situation, however. In an effort to achieve return, we cannot push rates out of the reach of consumers.

I wonder also if there is not a somewhat exaggerated emphasis on growth. We need to make the whole entity--the entire communication system--viable as well. Perhaps also 15% annual growth should be the long-term mark. We can do more than that for four or five years, but Mr. Stern is correct, beyond those years, foreign exchange requirements might impede such rapid growth.

MR. STERN:

I certainly agree that the spatial distribution of telephones is an important variable, particularly when we discuss rural systems. In pursuing the growth that I discussed earlier, I think it proper that countries allow urban systems to cross-subsidize the rural. The urban systems are expanding so quickly that such a policy could prove beneficial.

QUESTION:

How does the World Bank quantify the indirect effects of telecommunications on development?

MR. STERN:

At the World Bank, decisions about telecommunication loans are made independently of other projects in progress in a country. In addition, we take a conservative approach to estimating the returns for projects. Nevertheless, the internal rate of return of telecommunications projects is high, and there is a great amount of literature which discusses the broader economic returns of telecommunications investment. Perhaps at this point, we do not need more quantitative proof of these economic returns--we just need to keep repeating what we have already learned. What is more important is research about the foreign exchange effects of telecommunications investment. This is an area that the Bank would like to examine, and we would certainly welcome any analysis from those working outside of our organization.

QUESTION:

As we examine the problem of obtaining financing for developing countries' telecommunications systems, should we not also examine the other side of the exchange supply equation? In this way, we should note the effect of world commodity prices on these countries which must depend on raw material exports for foreign exchange.

MR. STERN:

I agree. Commodity prices and terms of trade have a far greater effect on a country's economic development than any aid program. For example, a four or five percent change in a commodity price can actually wipe out the effects of a large aid program. This sort of change caused problems for developing countries in the late 1970's, and in the 1980's this situation is compounded by significant real interest rates. It is difficult to address this global question at a conference such as this one. Within the telecommunications community, however, what we can do is to encourage competition among telecommunication suppliers in countries that may be willing to subsidize their telecommunications sectors.

QUESTION:

You have just described a major satellite project, with two satellites offering an enormous amount of capacity. What are your plans for using this capacity? Specifically, are there plans for sharing services with some of your neighboring countries?

DR. ALBERNAZ:

We are not currently making plans for sharing the satellite capacity, nor have our neighbors yet requested such arrangements. Should such an opportunity occur, however, we would be happy to discuss these ideas. In regard to the planned uses of capacity, we expect to increase the number of transmit stations from 24 to 40. Because we are currently renting time on INTELSAT's systems, there is not a powerful incentive for satellite use in Brazil. When the transponders from the Brazilian project because available, however, we expect new applications to emerge, particularly for rural communications.

QUESTION:

Please clarify the distinction you made between the need for telephone versus the demand.

DR. HUDSON:

I was pointing out that we ought to look beyond the demand for traffic. In rural and remote areas, there are few alternatives to telephone communication, and therefore behind the high traffic demand, there is a real set of communication needs. In addition, there is not the incentive of high return which results from high traffic demand. For example, in the South Pacific, in the outback of Australia, and in some parts of Africa, population density is quite low, yet alternative means of communication are few. In these places, rate of return calculations based solely upon revenue do not describe the entire situation. It is at this point that we need to look at the larger economic returns that Richard Stern was discussing earlier.

QUESTION:

When I consider the various developments taking place through the Maitland Commission, the ITU, and the OECD, it seems that the implementation of their plans for development and communications will take 100 years. What will we be discussing at conferences like these in 10 years?

MR. STERN:

I think that ten years will probably bring marginal improvements -- fairly boring prospects, perhaps. Yet I think that there is a new popular enthusiasm of telecommunications technology which may inspire faster progress than we might otherwise predict.

DR. PARKER:

One reason we're not seeing as much change as we'd like is that we still haven't addressed the institutional issues. We have assumed that urban and rural telecommunications are the same, that North and South technical needs are similar, and that cross-substation is the only way to fill the information gaps. Perhaps we need to change our assumptions, particularly in regard to institutional arrangements. For example, perhaps there is an alternative to a monopoly status of telecommunications in developing countries, so that the risks of telecommunications development are dispersed through the private sector.

QUESTION:

In making a case for investment in telecommunications in the developing world, what real evidence do we have about the connection of the investment to development? Specifically, what have we done to analyze the effects of systems that are currently in place?

DR. HUDSON:

There have been some excellent studies completed in the past few years, so that we now have a critical mass of literature. Still, more studies are needed. Many of the micro case studies have been useful, but we are limited in what we can extrapolate from them. We need to approach the problem both theoretically and practically. In a time when the world economy will not allow blanket spending for telecommunications projects -- however badly they are needed -- we must help planners get more efficient and appropriate use of the telecommunications technologies that they can acquire.

I think that further technological innovation will speed our progress in this regard. I also have faith that once a system is in place, the consumers themselves will demand faster progress. Once Australia's satellite is deployed, for example, I expect many groups will come forward to demand services.

PANEL 5: FUTURE DIRECTIONS IN SATELLITE POLICY

The final panel of the conference addressed current issues in international satellite policy, focusing on the World Administrative Radio Conference on the Use of the Geostationary Satellite Orbit and the Planning of the Space Services Utilizing It (the Space WARC or ORB 85), the first session of which will be held in Geneva in August 1985. This conference will examine techniques to "guarantee in practice for all countries equitable access to the geostationary satellite orbit and the frequency bands allocated to space services." The reason for leaving the discussion of satellite policy issues until the last session was to enable the participants to relate the information on available satellite technologies, needs of developing countries, and plans for satellite systems to serve developing countries to the discussion and analysis of policy issues.

The first presentation on "Issues for WARC (ORB) 85 and 88: Some Perceptions" by T.V. Srirangan, was actually the second part of his paper "Why Orbit Planning: A View from a Third World Country." The first part, on "The Indian Experience in Satellite Applications," was presented in Panel 3. As previously noted, Mr. Srirangan's career includes representation of India at the ITU, where he has been Councillor from India on the ITU Administrative Council, and has led several Indian delegations to several ITU meetings and conferences, where he chaired several committees and subcomittees. He also chaired the WARC on Aeronautical Mobile Services in 1978. At the 1979 WARC, he played a key role in the deliberations which resulted in Resolution 2, which mandated holding the Space WARC.

The second speaker was Mr. Donald Tice, senior diplomatic officer on the U.S. Space WARC delegation. Mr. Tice is a career member of the U.S. government's Senior Foreign Service, and is currently a member of the staff of the Department of State's Office of National Communication and Information Policy. Mr. Tice's previous diplomatic experience includes acting as Executive Secretary of the U.S. Delegation to the Strategic Arms Reduction Talks, and tours for the Department of State in Yugoslavia, Bulgaria, Belgium, and Canada. His paper examined U.S. policies in communications and their relationship to the Space WARC.

The final speaker was Dr. George Codding, Professor of Political Science at the University of Colorado at Boulder. Dr. Codding is the author of two major books on the ITU: *The ITU: An Experiment in International Cooperation* (1952) and *The ITU in a Changing World*(1982), the latter co-authored with Anthony Rutkowski. He has carried out several studies of ITU conferences for the International Institute of Communications, and is a member of the board of the University of Colorado's Masters Program in Telecommunications. His paper proposed several measures designed to build international confidence in the ITU.

This final panel was followed by a discussion session which involved speakers from previous panels and participation from the audience. A summary of this session follows the papers.

WHY ORBIT PLANNING: A VIEW FROM A THIRD WORLD COUNTRY
PART II - ISSUES FOR WARC (ORB) 85 & 88: SOME PERCEPTIONS

by

T.V. Srirangan

Member (Telecom. Development) P&T Board and

Ex-officio Additional Secretary

to the Government of India

Ministry of Communications

SOME FACTS

It is useful to recaptulate certain well-known facts. The Geostationary Orbit (GSO) is a unique and finite natural resource. Satellites located in this orbit offer many advantages. As with all manmade satellites, their control and operation are critically dependent on the use of the Radio Frequency Spectrum (RFS). The location of any satellite in the orbit is heavily influenced by consideration of mutual RF interference effects with respect to other satellites in the vicinity, which are also using the same operating frequency bands, in order to ensure efficient operations of all the satellites involved. The "crowding" or "congestion" of the orbit, which one hears of so often these days, should be understood in terms of the limitations imposed on the maximum number of satellites in specific frequency bands which can be located in any given segment of the orbit without causing interferences beyond an acceptable limit. The orbit and the frequency spectrum are thus inextricably linked. The congestion is not to be viewed in terms of physical occupancy of the orbit by satellites or the fears of mutual collision. That situation is not yet upon us.

The Radio Regulations (RR) of the ITU, which provides the international framework for the efficient and satisfactory use of the RFS resource, allocates specific bands or parts of the spectrum for various space telecommunication services. A majority of the allocations are on a shared basis for many space and terrestrial communication services. A complex set of technical and regulatory provisions in the RR are designed to ensure that operation of all RFS based services conform to those provisions. Otherwise, there would be utter chaos in the spectrum. Historically, the regulatory mechanism has evolved from the principle that an existing and registered operation is entitled to full protection against harmful interference from any subsequent operation. This appeared fair enough in earlier years. But with technological strides and the increasing variety and applications of systems derived from the RFS, and the growing demands for related services all over the world, the "first come, first served" basis of the RR was no longer adequate for the exploitation of a limited natural resource which all countries were entitled to share equitably. Progressive changes to the RR had to be adopted in consequence, particularly since the 1950's.

International sharing of the spectrum does not pose problems in respect of all frequency bands and services. Essentially, where the radiowaves can propagate over considerable distances and cause interference, the problem becomes severe. For terrestrial services, this is true mostly for services operating in the part of the RFS below about 30 MHz. Above that threshold, signal propagation is mostly on line-of-sight basis, and in terms of interference potential, is largely limited to the area determined by the radio horizon. This generally calls for coordination between countries only in border areas and does not usually present insurmountable problems. That is why the thrust of the "planned approach" to the use of the RFS was mainly confined initially to the terrestrial services using the long, medium and short wave bands below 30 MHz.

The advent of satellites altered that situation since, inherently, the radio horizon for a satellite-based transmitter or receiver is determined by the altitude of the satellite above the earth's surface. For a satellite located in the GSO, the horizon encompasses more than a third of the earth's surface. Thus, there has been a sea change in the interference range and environment, which attracts a host of space and terrestrial services. In this scenario and in the context of rapidly rising worldwide demands for satellite based services, it was therefore natural that the desire to ensure a planned exploitation of the inseparable orbit frequency resources was manifested in forceful terms.

As in the case of several others services, space telecommunication services have also been regulated on the principle of right of protection to existing operations. In other words, any later entrant has the onus to show that his proposal does not adversely affect earlier operations. Without the "consent" or "coordination" of the latter, the former cannot "enter" the orbit, unless he is prepared to modify his proposals suitably. This coordination process, fairly well articulated in the RR, is in a bilateral mode, with each affected space network/administration and often becomes highly interrelated, and hence iterative in practice, apart from being intricate. There are technical norms, no doubt, but success is largely governed by "good will". There are no mandatory provisions to compel all concerned parties in a given case, to conform to a common, equitable norm. That is where it hurts the nations of the developing world, which for reasons of history, technology gaps and paucity of resources are always "later entrants." They have, under the present dispensation, to trail behind perpetually or accept penalties to derive the benefits of satellite which they are supposed to share with equal rights.

The principle of equitable sharing of the frequency spectrum for space communications by all nations was adopted by the EARC in 1963, the year of SYNCOM. This has been reiterated and further strengthened by successive ITU conferences. A Resolution exists, as a part of the RR, specifically laying down that there could be no permanent claim by any administration for any orbit position. Further, ITU's 1973 (Malaga) Convention 131 states:

> "In using frequency bands for space radio services Members shall bear in mind that radio frequencies and the geostationary satellite orbit are limited natural resources, that they must be used efficiently and economically so that countries or group of countries may have equitable access to both in conformity with the provisions of the Radio Regulations according to their needs and the technical facilities at their disposal."

This underwent a change at the Nairobi (1982) Plenipotentiary Conference. Article 154 of the Nairobi Convention states:

> In using frequency bands for space radio services Members shall bear in mind that radio frequencies and the geostationary satellite orbit are limited natural resources and that they must be used efficiently and economically, in conformity with the provisions of the Radio Regulations, so that countries or groups of countries may have equitable access to both, taking into account the special needs of the developing countries and the geographical situation of particular countries.

The Nairobi modification spotlights the concerns of the developing world. Despite all these principles, enshrined in the basic instrument of the ITU, paradoxically the actual regulatory procedures have continued to be structured essentially on the foundation of right of protection to existing operations, an euphemism for "first come, first served." The Broadcast Satellite Service in the 12 GHz band, which is subject to an orbit-frequency assignment plan, is a notable exception.

NEED FOR CHANGE

The WARC-79 in adopting Resolution 3, recognized the inherent contradiction within the ITU Convention and the RR, and called on the world community to embark on the planned utilization of the orbit spectrum resources with a view to guaranteeing, in practice, equitable access to the same by all countries. The WARC (ORB) 1985 and 1988 will address this matter. Resolution 3 clearly recognized that the utilization of the GSO was heading towards congestion, and corrective measures were urgently needed. Though the fears of congestion were hotly contested at WARC-79, there is now wider acceptance of its validity.

SOME BASIC PREMISES

There is no universally accepted definition of "equity". No doubt, it cannot mean "equal" sharing of the resources by all countries, big and small, even though they are all equal sovereign states. The dictionary meaning of "equity" being "fair" or "just", a wide enough interpretation is a prerequisite for the launching of a new, egalitarian regime for regulating the exploitation of these resources. A corollary would be the adoption of a suitable yardstick to verify the projected needs of countries, lest genuine requirements of one or the other are cast aside.

By "guaranteed access in practice", it should be understood that irrespective of the chronological sequence of "entries" into the orbit in any frequency band, a "later entrant" shall not be discriminated against in any way. He shall "enter", and if there are adjustments or penalities resulting from the same, all concerned parties shall bear the burden equally. Mandatory provisions in this respect should be the essence of any new regime.

There is need for a clearer perception of what efficient use of a resource connotes. Irrespective of what technology can make possible for more "efficient use", in the situation of gaping disparities in the world of today in terms of economic and technological development, the dominant thought should be of "efficacious" use. The translation of orbit efficiency into "maximum equivalent speech channel capacity" is neither correct nor wise. It will not carry much meaning or purpose to the developing world. The Nairobi Convention Article 154 puts this succinctly.

Telecommunication services are now recognized as a basic infrastructure for national development. They need, relatively, a far greater emphasis in the developing world, to promote accelerated national growth to make up for the omissions of the past. The medium of geostationary satellites has dramatic and demonstrated potential to do just that. The future regime should facilitate translation of that potential to practical benefits.

APPROACH TO PLANNING

Of the various space telecommunication services defined in the RR, the immediate problems of congestion and inequitable sharing related mainly to the Fixed Satellite Services, particularly, in the frequency bands below about 15 GHz. These bands have to accommodate the rapid growth of international, regional and domestic satellite networks all over the world. The developing countries, have, as yet, a relatively minor presence, but their requirements are set for substantial future growth, especially in the 6/4 GHz band which is advantageous from the points of technology capabilities, economics and propagation. A majority of developing countries are located in the tropical zone, where higher frequency bands could be subject to propagation problems.

In the years since WARC-79 several approaches to orbit planning have been proposed and debated. Without going into the details of their relative merits and demerits, it can be said that neither cosmetic changes to regulatory procedures nor the adoption of a so-called, a-priori, long term, orbit-frequency assignment plan, is likely to find a consensus. As mentioned earlier, mandatory provisions, in the form of an internationally accepted legal instrument to enforce equitable and guaranteed access would carry conviction to the developing world. In the ITU, hitherto, assignment plans of one kind or the other would have fulfilled such an objective.

A REASONABLE MIDDLE PATH

In my view, adoption of an assignment planning cycle of 8-10 years, which is of the same order as for the planning and implementation of a satellite network, inclusive of the design life span of the satellite, would offer the best prospect for wide endorsement. Over such a time frame realistic need projections can be expected and possibly also validated, so as to set at rest fears of exaggerated projections which could negate optimum use of the orbit. There is no long term reservation of orbit spectrum and the consequent wastage of resource, to no one's benefit. No undue technology freeze will result. A one time planned accommodation of all existing and projected valid needs is the best way of ensuring the most efficacious use of the resource. Adjustments and penalities can be equally distributed among all networks, without scope for discrimination. A reasonable state of stability of operation of all networks would be possible over the planned period. That is something inconceivable in any other step-by-step or periodical coordination process.

In such an approach, for the first planning cycle, it is recognized that special transitional arrangements would be necessary in respect of existing or "ready for operation" networks, since they may not be able to conform fully to the technical provisions of the plan. However, such satellites would be due for replacement during the second planning cycle, and could then be called upon for full conformity. It should also be noted that an 8 to 10 year cycle would not ordinarily necessitate latitude or flexibility to provide for unforeseen requirements. Geostationary satellites cannot be planned, built and launched in "quick time", not even by the few powers who have total capability in this regard, and they, in fact, are in a position also to foresee more clearly their needs. Nevertheless, some provisions to take care of genuine emergent situations should be possible.

SOME RELATED ASPECTS

One cannot but take note of arguments which continue to be advanced that future needs of developing world can be met by technology advances. This shows inadequate awareness of the technology situation in the developing world and of their aspirations for achieving a substantial measure of self-reliance in technology, either individually or collectively. Developing countries attach great value to this, for they have learned their lesson from their own history and do not wish to be subjugated by yet another kind of dependence. There are also overtones of national pride and prestige, and this is true of <u>all</u> countries. The "technology advances" premise does not respond to these concerns.

Mention is also made in this context of solutions to orbit congestion being found through satellite clusters, and antenna farms. They are becoming technologically feasible but are certainly not around the corner, and quite some distance away, as operational propositions. Even when they become so, there remains the crucial question of who controls them? This brings us back to the cardinal issue referred to in the preceding paragraph.

CONCLUSION

With the first session of the Space WARC (ORB 85) less than a year away, there is need for understanding of the needs and concerns of the developing world, so as to pave the way for a better future for all. I trust this presentation will assist this process in some small way.

REFERENCES

Srirangan, T.V. "The INSAT System - An Overview." Seminar of Satellite Communications. Advanced Level Telecom Training Centre, Ghaziabad, India, March 1978.

Srirangan, T.V. "INSAT-1 Domestic Satellite System for India." ITU Regional Conference-cum-Seminar on Development and Management of Telecommunication, Bangkok, January 1982.

Srirangan, T.V. "INSAT-1 -- The Multipurpose Domestic Satellite System for India." ITU (CCIR) Seminar, Shanghai, October 1983.

Srirangan, T.V. et al. "INSAT -- The Indian National Satellite System -- A Case Study." ITU (CCIR) Seminar, Shanghai, October 1983.

Srirangan, T.V. "Some Thoughts on Techno-Economic Considerations and Potentials of Orbit/Spectrum Planning for Developing Countries." International Training Course on Orbit-Frequency Planning, Space Application Centre, Ahmedabad, February-March 1981.

Srirangan, T.V. "Equity in Orbit - Planned Use of a Unique Resource." International Institute of Communications, Annual Conference, Berlin, September 1984.

ISSUES IN U.S. INTERNATIONAL TELECOMMUNICATIONS POLICY

by

Donald C. Tice

Deputy Director, U.S. Delegation to the Space WARC

Department of State

Washington, D.C.

I would like at the outset to thank the College of Communication and Heather Hudson for inviting me to participate in this meeting as a representative of the Department of State's Coordinator for International Information and Communication Policy.

What I would like to do is briefly mention some of the broader United States policy objectives in the area of communications and information, review what we see as some of the future trends in the field, and relate these to the Space WARC.

The United States has a number of broad objectives which it seeks to serve through its policy in international telecommunications and information. These include:

-- to enhance the free flow of information and ideas among nations;

-- To promote, in cooperation with other nations, the development of efficient, innovative and cost-effective international communications services responsive to the needs of users and supportive of the expanding requirements of commerce and trade;

-- To ensure efficient utilization of the geostationary orbit and electromagnetic frequency spectrum;

-- To expand information access and communications capabilities of developing countries to facilitate their economic development;

-- To ensure the flexibility and continuity of communications and information required to maintain national defense and international peace and security;

-- To promote competition and reliance on market mechanisms which assure efficient prices, quality of services, and efficient utilization of resources; and

-- To promote the continuing evolution of an international system of communication services that can meet the needs of all nations of the world, with attention directed particularly towards providing such services to economically less developed countries.

The common purpose of these goals is to promote free flow of information between nations and people of the world. But they are pursued in a dynamic framework in which the pace of technological change makes it at times difficult to predict with accuracy what will be, to use a term common in ITU circles, the situation prevailing at a given point in the future.

This pace will pick up in the coming years, as mutually reinforcing advances in technologies not only improve existing capabilities but open new fields of activities. Some of the basic characteristics of the trends in these advances can be described, however, and they can help in understanding potential future effects. Some of these characteristics are:

Decline in cost of information transfer. Continued reduction in unit cost of computation and information processing (by several orders of magnitude in recent years), increase use of digital signal processing, and broadband transmission techniques continue to reduce the unit cost of transferring information. Physical limits to substantial further reductions have not yet been approached.

Further miniaturization, more reliability, greater performance. Increases in circuit density of solid state electronics unimaginable a few years ago. Today, for example, a five-inch diameter wafer of integrated circuits contains 20 million transistors. This has reduced the cost of manufacture, operation and maintenance. The cost of an integrated circuit in 1967 was 34 dollars; today it is 67 cents. These characteristics contribute to the wider dissemination of these technologies and their devices throughout the world.

Advances in terrestrial communication capabilities. While one cannot with complete accuracy anticipate the effect of such advances, it is clear that the added terrestrial capacity provided by the installation of fiber optic systems will have an effect on the use of space-based communication links. At the least, availability of fiber optics on the principal transoceanic routes will tend to ease congestion in the GSO. Some even see an eventual division where satellites primarily will service mobile customers and thin route links while optic fibers will primarily service stationary customers and high usage links.

These, obviously, are only a few of the trends which are emerging in the area of telecommunications and information. I believe it can be safely said that by the end of the 20th century, communications and computer technology will have made information and its management one of the most important factors in international economic activity.

A significant portion of the technology supporting information flows ultimately rests on the ability to assure the orderly use of the radio frequency spectrum, the natural resource on which all radio communication is based, and, in the case of satellite communications, on efficient utilization of the resource represented by the geostationary orbit. In the absence of a globally accepted allocation of spectrum to support each radio communication service, and a regime for harmonizing the assignment of frequencies within each allocation, there would be harmful interference between the radio services of different countries, critically diminishing the value of the spectrum resource.

This, of course, highlights the importance of the International Telecommunication Union. The United States recognizes that its interests and those of other member administrations have generally been well served by the ITU through its internationally accepted mechanisms for global allocation of spectrum and harmonization of frequency assignments. There have been those in the United States who have favored the establishment of some alternative mechanisms which would produce, in their view, a superior forum for negotiation and accommodation. This is because a widespread consensus exists that there must be order in the use of the spectrum, and broadly accepted criteria for the design and operation of international telecommunications networks.

The challenge to the United States and to the ITU membership as a whole is to strike a balance between national interests and the collaboration that is prerequisite to the fullest development of telecommunications technology and services internationally.

This, then, brings us to the Space WARC and the very real and difficult questions that it is to address.

Over the past twenty years, the pace of advances in communications satellite technology has been such that the available capacity of the GSO continues to exceed the demand for services. And, this growth has been accomplished with an efficiency which has kept the cost of these services, measured in terms of cost per communication channel, well below projections based on inflationary trends during this same period.

Nonetheless, the accelerating pace of development activity and implementation of systems using the GSO has led to concern on the part of a number of countries that when the time comes that they wish to avail themselves of the GSO resource, that resource will not be readily available to them. The Space WARC first session, next summer, is to address how the world community can provide equitable access for all countries to the GSO and related services.

I will not go into detail on the various issues which arise in this context, because most of you here are already conversant with them. What I do want to do, however, is to state clearly the U.S. objectives in the coming conference and review the State Department's preparatory activities.

First, we strongly believe that an outcome at the conference that introduces a degree of rigidity in governing use of the GSO which slows the pace of investment and research into expanding the capacity of the GSO would be contrary to the interests of everyone and every nation, developed and developing.

Second, we believe just as strongly that the conference should arrive at an outcome which provides a degree of guaranteed access to the GSO which is acceptable to a large majority of the ITU member administrations.

We do not believe that these two outcomes are mutually incompatible, but in order to accomplish both of them, solid preparation will be required on our part and on the part of all other administrations.

When we returned from the Conference Preparatory Meeting (CPM) for the Space WARC, held this past summer in Geneva under CCIR auspices, one of the first things we did was review how we could best use the resources and knowledge represented by the private sector in the preparations for the 1985 conference. This led us to review and then revise the work program for the FCC's Space WARC Advisory Committee, placing focus on the areas where we knew from the CPM that we needed more study and analysis, and shortening deadlines so that we would have the benefit of the Advisory Committee's work for use in our policy process.

The we set up a small, core Space WARC group, including representatives from the Department of State, the Department of Commerce, the Department of Defense, and the Federal Communications Commission, and chaired by our designated chairman for the 1985 conference, the Honorable Dean Burch. We also included representatives of the private sector who can act as a continuing, real-time bridge between the work of the core group and the Advisory Committee.

In our work, we also are drawing on the results of studies undertaken by the Commerce Department/NTIA group on Space WARC, Ad Hoc 178, established under the IRAC framework.

Finally, we have continued to hold monthly meetings of the Interagency Steering Group for Space WARC, which draws together all of the government agencies and entities concerned with Space WARC policy. This Committee is the mechanism both for disseminating information within the government about ongoing activities, and for drawing on the resources of agencies as diverse as NASA and USIA for studies and other advisory assistance.

Further, in order to inform ourselves on the thinking in a wide spectrum of countries, we have sent messages to our embassies abroad briefing them on our preparations to date and asking them to consult with their host governments and provide us with the benefit of those consultations. In addition, we have already undertaken consultations with key countries to learn how they are approaching preparations for WARC 85, and to share our thinking and approaches.

After the first of the year, when our policy process is further advanced, we plan to undertake a broad round of consultations with some 35 countries, and to attend all three of the regional seminars which the ITU will sponsor.

I would not want to leave the impression that all of these activities have as yet resulted in a firm U.S. position paper for Space WARC with all the "t's" crossed and all the "i's" dotted. To the contrary, it is a dynamic process, and there are still many differences in views both within the government and the private sector, and between the two, which need to be resolved.

I believe I can say with confidence, however, that there is broad consensus on the two objectives I outlined earlier, that is, that WARC 85 must point the way for the 1988 conference to be able to provide an acceptable guarantee of access to the GSO for all countries, while retaining the flexibility which will allow the continued introduction of new, more efficient, and more economic technology.

CONFIDENCE-BUILDING AND THE 1985 SPACE WARC*

by

Dr. George A. Codding, Jr.

Department of Political Science

University of Colorado

The International Telecommunication Union (ITU) has scheduled a conference on satellite communications, the first session of which will be held in Geneva, Switzerland, in 1985 and the second in 1988, in order "to guarantee in practice for all countries equitable access to the geostationary-satellite orbit and the frequency bands allocated to space services."(1) This conference provides an excellent opportunity for the nations of the world to take the necessary actions that will assure the continued orderly progression of the use of this extremely valuable communication tool.(2)

Unfortunately, a controversy over the proper meaning of the term "equitable access" has arisen that poses a serious threat to the success of this conference. Basically, the debate on this question has boiled down to two seemingly irreconcilable positions. On the one hand, there are those

* The author wishes to thank the University of Colorado Master of Science in Telecommunications Program for its financial assistance and Kathleen Blue, a graduate student in the University of Colorado M.A. in Political Science Program, for her research help.

who believe that it is necessary to allot specific geostationary satellite orbital positions and the necessary radio frequency channels to designated countries to be used whenever they are in a position to do so, known as the "*a priori*" approach.(3) On the other hand there are those who are of the opinion that the needs of both the present and future users of geostationary satellite communications can best be met by a more flexible approach. This approach depends largely on technological advances and new ways of employing satellite communications to guarantee to the newcomers that they will find space in the geostationary orbit and the appropriate frequencies when the time comes that they are in a position to utilize them. The first position is the one advocated by a majority of the members of the ITU, including most of the developing countries; while the second is the one taken by some of the other members whose main spokesman has been the United States.

The Third World has become convinced of two things. First, satellite communication will aid them immeasurably in their development. Second, because of the heavy actual and planned use of the geostationary satellite orbit by the United States and other developed countries, when the time comes that they have the necessary resources to take advantage of communication by this means, there may be no appropriate orbital positions or the necessary frequency channels left for them to use. It thus may be necessary, so the argument goes, to allot the available orbital positions and frequency channels to all of the countries planning to use them. As stated by the delegate from China to the ITU's 1979 World Administrative Radio Conference:

> For many years, facts have shown that because of the imbalance between economic and technical development in the various countries, the pertinent provisions of the present Radio Regulations cannot guarantee the right for all countries to use the orbit and the frequencies on an equal footing. This is because the existing provisions are based on the principle of 'first-come, first-served.' As long as this principle prevails, the orbit and frequencies will be occupied first of all by those countries who got there first. Whereas, the developing countries who came after, missed their turn in orbit. We wish to break away from the principle of 'first-come, first-served.' There is only one means: to replace it by a plan. A plan would permit ensuring that each country has equal rights with all others. So that when it finds it necessary to do so, it can occupy the place which is its by right. (4)

The United States has argued that the true benefits of geostationary satellite communication can be reaped only if the technology and the uses of that technology are allowed to evolve on their own terms. Flexibility is an essential ingredient in that evolution. The *a priori* planning method, according to the United States, is by its very nature inflexible and thus would hinder that development. It is also useless since most of the world will not be able to take advantage of this type of communication for decades to come, if ever. Furthermore, if satellite communication is allowed to evolve, it is very likely that it will result in a much more efficient use of the satellite orbit and frequency channels; thus there will be adequate space in the future for any country wishing to take advantage of this type of communication. This is clear in the report of the United States delegation to the 1971 World Administrative Radio Conference for Space Communications. The U.S. expressed its opposition to holding of conference in 1977 to create an *a priori* plan mainly because:

> ...such a conference would likely, and largely on the basis of unvalidated requirements, put 'country name tags' on radio frequency channels and geostationary orbit positions, thus severely restricting the effective use of those natural resources in what could be envisaged as a slow by evolutionary development of broadcasting satellite systems. (5)

Some background concerning the manner in which the ITU treats terrestrial radio communication is necessary at this point in order to understand the nature of this dispute. When radio communication became operational near the beginning of the century, a practice developed of notifying the Secretariat of the ITU of frequency assignments that member administrations wanted to have protected from harmful interference by the radio stations operating under the authority of other member administrations. Countries soon began using the resulting notification list as a guide in the choice of frequencies for new radio services, thus in effect giving the first in time a priority over all succeeding notifications.

This practice, later dubbed "first-come, first-served," was not challenged until the ITU's Atlantic City Conferences in 1947. At that conference, the United States proposed that the list of notified frequencies by completely revised in order to reflect the actual and future needs of all of the members of the ITU. Once that was done, a new international body would be created within the ITU to approve additions or changes to that list, and to settle any dispute that should subsequently arise. The new body, the International Frequency Registration Board (IFRB), was created (6) but, because of international politics and a fear by some countries that the process would impinge on national sovereignty, it proved impossible to draw up the new "engineered" international frequency list.(7)

The IFRB was then downgraded to little more than a recorder of frequency assignments. In essence the procedure that resulted is as follows. Administrations notify the IFRB of frequency assignments, or major changes in the characteristics of existing assignments if the use of that frequency is capable of causing harmful interference with the service of another administration, if it is to be used for international communication or if the administration wants international recognition of the use of that frequency. The Board makes an examination of the frequency assignment to determine if it will cause harmful interference and if it is in conformity with the provisions of the ITU's Radio Regulations. If the Board's findings are favorable, it enters the assignment in the Master Frequency List. If not, the assignment is returned to the administration that submitted it. Normally, the administration in question will find an alternative frequency or make the necessary changes to make it conform to the provisions of the Radio Regulations. If an administration insists, however, an assignment which has been given an unfavorable report by the Board must be registered in the List if it can be shown that it can be used without causing harmful interference. The Board also has the right to review assignments for actual usage and can cancel those which are not in use if the administration so agrees.(8)

Because of the costs involved in maintaining the Board and the minimal nature of its duties, a number of proposals were introduced at the ITU's 1965 Montreux Plenipotentiary Conference to abolish the IFRB and turn its duties over to the ITU's Secretariat. Although the number of its members was cut from eleven to five, the Board was saved from extinction by the new developing country members of the ITU who saw the Board as a friend in court and an important source of advice on frequency management problems.

When the ITU first became involved with satellite communications, it established a procedure similar to the one that it had previously adopted for radio frequencies.(9) Member administrations are required to notify the IFRB of the details concerning any plans to initiate a new geostationary satellite communication service. The Board publishes that information and circulates it among the members of the ITU. If, after studying that information, another member administration is of the opinion that one of its existing or planned satellite services would be adversely affected by the new service, the parties involved are requested to settle their differences. If they are unsuccessful in that endeavor, the parties may seek the assistance of the Board. The Board then acts as a facilitator for the parties to the dispute in their efforts to find a solution to their problem. There is, however, no obligation on the parties to seek the help of the board or if they do, to take its advice. Most importantly, there is no obligation on the part of the party whose notification is objected to, to make changes in its notification. This procedure, therefore, is seen by the developing countries as giving the first administration to start a satellite service an important advantage over those which come later.

The concerns of the developing countries over the fairness of this system led to the insertion in the International Telecommunication Convention, the ITU's basic treaty, of the following provision:

> In using frequency bands for space radio services, Members shall bear in mind that radio frequencies and the geostationary satellite orbit are limited natural resources and that they must be used efficiently and economically, in conformity with the provisions of the Radio Regulations, so that countries or groups of countries may have equitable access to both, taking into account the special needs of the developing countries and the geographical situation of particular countries. (10) (emphasis added)

It also resulted in the calling of a conference which created an *a priori* allotment plan for geostationary satellite broadcasting in the 12GHz band for Regions 1 and 3 in 1977 and for Region 2 in 1983, over the strenuous objections of the United States (11), and in the end it resulted in the decision to hold the conferences on geostationary satellite communication scheduled for 1985 and 1988 that we are discussing here.

It must be admitted that in the past few years there has been a certain amount of softening in the two opposing positions that were described earlier. The United States, for instance, found that the plan created by the 1983 Region 2 Broadcast Satellite Conference was, in general, satisfactory, despite its earlier objections.(12) And, in a recent Notice of Inquiry, the Federal Communications Commission suggested that it would not rule out the possibility of the 1985/88 conference establishing an *a priori* plan for satellite communication as long as that plan is not overly "detailed" and too "inflexible".(13)

There has also been a softening of the position of some of the leaders of the Third World. Both India and China, which are becoming important users of satellite communications, have made it known recently that under the proper circumstances and with the proper guarantees, they would be willing to explore the possibility of supporting a flexible and not too detailed _a priori_ plan at the upcoming satellite conferences.(14)

The organizers of the 1985 and 1988 conferences would be foolish, however, if they were to proceed on the basis that the issue of equitable access has been laid to rest. There are numerous details, both major and minor, that will have to be settled before even the positions of the major spokesmen for the two sides of the issue can be said to provide a genuine basis for compromise. Most important of all, the countries that have been thus far involved in this dialogue are only a small minority of those which will be involved in the two conferences. The vast majority of the participants will come from countries that are much further down the road as concerns having their own geostationary satellite communications networks. Although the rigid, treaty-based _a priori_ approach may have its drawbacks, including the possibility of somewhat retarding the development of satellite communications, they may well consider it a cheap price to pay for the guarantee of future access to geostationary satellite communication that this approach would provide. Since these countries will have a majority of the votes, it is essential that something be done to allay their fears if the conferences are to have a chance at success.

I would like to suggest three steps to achieve this goal. All three have as their purpose the creation of an atmosphere which will assure the developing world that it will indeed have access to the geostationary orbit and the necessary frequency channels when the time comes that they can take advantage of them. All three steps involve the International Telecommunications Union.

Step one would consist of making a substantial U.S. contribution to the ITU's Special Voluntary Program for Technical Cooperation, the successor to the ill-fated Special Fund for Technical Cooperation (15) to be used to help the poorest of the developing countries or groups of countries to plan appropriate telecommunication networks, including the use of geostationary satellites. This could take the form of sending experts to the developing countries or sending individuals from the developing countries to the more technologically advanced countries for training.

Too often the developing countries do not have the necessary trained personnel. If help were made available to overcome this serious deficiency, it could be expected that the poorest of the developing countries would be in a better position to create appropriate telecommunication systems, including a realistic use of geostationary orbital communications. Most importantly, it would help create a sense of confidence in the future which would make those countries less in a hurry to insist on a rigid *a priori* allotment plan.

Step one would necessitate a major change in policy on the part of the United States, but it would not be much of a financial burden, since a large portion of the contribution would probably be returned to the United States in its roles as major supplier of experts and trainer of students in modern telecommunication techniques.

The second step would be to increase the ability of the ITU's organs to help the developing countries plan and operate satellite communication systems. This would include the creation of a special Working Group of the International Radio Consultative Committee to deal exclusively with satellite communication systems in the developing world and hiring one or two additional engineers with special expertise in satellite communication for the Group of Engineers presently working in the ITU's General Secretariat. The Secretary General could also be directed to make an extra effort to provide the developing countries with the information that they would need in their preparations.

There would also be financial costs involved in implementing step two, but they would not be of a major nature. In any case these costs would be much less than those which would be involved in other proposals that have been made by delegates from developing countries to ITU conferences in the past, such as those for the creation of a new international consultative committee devoted exclusively to problems of development.(16)

The most important step, however, and the one I want to explore in some detail here, is the upgrading of the role of the ITU's International Frequency Registration Board (IFRB). As mentioned earlier, at the present time, the IFRB's main task as regards satellite communications is to keep a register of notifications of new communication satellite services for which administrations want protection. If an administration complains about the effects of a new satellite communications service, the Board's role is confined primarily to acting as facilitator, informing the parties of the need for coordination, requesting information, and the like.

It is proposed that the IFRB be elevated from facilitator to mediator. The IFRB should be involved in every stage of the process of activating a new satellite service. It should be permitted to offer its advice, on its own authority, to a new entrant on all aspects of its proposed service, including the appropriate orbital position, frequency channels, power, and antenna design. Where there are complaints that a new service might conflict in any way with an existing or already planned system, or where the Board itself feels that there will be such a conflict, the Board should have the authority to suggest solutions to all parties concerned. Most importantly, this would include requests to existing services to make adjustments that would permit a new satellite network to begin service without undue penalty.(17)

This change could be instituted easily by the 1985 Space WARC under the present provisions of Article 10 of the ITU Convention which lists among the Board's essential duties, in addition to the registration of frequency assignments and positions assigned by member countries to geostationary satellites:

> To furnish advice to Members with a view to the operation of the maximum practicable number of radio channels in those portions of the spectrum where harmful interference may occur, and with a view to the equitable, effective and economical use of the geostationary satellite orbit; [and]
>
> To perform any additional duties, concerned with the assignment and utilization of frequencies and with the utilization of the geostationary orbit, in accordance with the procedures provided for in the Radio Regulations, and as prescribed by a competent conference of the Union....

The proposal has the merit of building on existing foundations. A five member Board is already in place with a trained staff to support it. The members of the Board are chosen by the ITU's plenipotentiary conference in such a way as "to ensure equitable distribution amongst the regions of the world" from individuals who are "thoroughly qualified by technical training in the field of radio," and who "possess practical experience in the assignment and utilization of frequencies." (18)

The composition of the present Board is as follows:

1. Region A (The Americas): Gary C. Brooks (Canada);
2. Region B (Western Europe): William H. Bellchambers (United Kingdom);
3. Region C (Eastern Europe): P. Kurakov (U.S.S.R.)
4. Region D (Africa): Abderrazak Berrada (Morocco);
5. Region E (Asia and Australasia): Yoshitaka Kurihara (Japan).(19)

Since the Board was expected to have a great deal of authority, those who were responsible for its creation in 1947 surrounded it with a number of provisions that they felt would protect it from undue influence. For instance, the members serve "not as representing their respective countries, or of a region, but as custodians of an international public trust." In addition, a member of the IFRB is forbidden to "request or receive instructions relating to the exercise of his duties from any government or a member thereof, or from any public or private organization or person." Member countries, for their part, are required to "respect the international character of the Board and of the duties of its members and shall refrain from any attempt to influence any of them in the exercise of their duties." Further, administrations are requested to refrain from recalling their members during their term of office.(20)

The purpose of the proposed change in the duties of the IFRB, as with the other changes already outlined, is to give the developing countries confidence that they will indeed be in a position to take advantage of geostationary satellite communication in the future without the need for an inflexible *a priori* allotment plan. The developing countries already look on the Board as a friend to their cause. Having a friendly and competent adviser in court would make it easier for them to accept the assurances of the United States that new developments will indeed open up enough orbital positions and frequencies for everyone.

From the perspective of the developed nations, although it would be difficult to ignore the Board's advice under normal circumstances, since mediation is not binding, a government could refuse to comply if it felt that it would result in serious harm to its vital national interests.

Whether these changes are accepted or not will depend on the goodwill of the delegates to the 1985 Space WARC and whether they are truly interested in making the conference a success.

FOOTNOTES

1. See ITU, Radio Regulations, Edition of 1982, Resolution No. 3.

2. The full title of this conference is the World Administrative Radio Conference on the Use of the Geostationary Satellite Orbit and the Planning of Space Services Utilizing It. In this text we will use the more user friendly term "1985 Space WARC."

3. According to the terminology adopted by the ITU, frequency channels or satellite orbital positions are "alloted" to specific countries or groups of countries. Administrations "assign" frequencies to specific national users and the ITU "allocates" frequency bands to terrestrial or space radio services. See Radio Regulations, *op. cit.* paras. 7, 8 and 9.

4. As quoted in Heather E. Hudson, "Developing Country Orbit/Spectrum Interests: An Analytical Framework," a paper presented at the IIC Annual Conference, Berlin, September 22, 1984, p. 4.

5. See U.S., Department of State, Office of Telecommunications, Report of the United States Delegation to the World Administrative Radio Conference

for Space Telecommunications, Geneva, Switzerland, June 7 - July 17, 1971, TD Serial No. 26, Washington, D.C., August 16, 1971, p. 48.

6. The ITU has three permanent organs, the General Secretariat, the International Frequency Registration Board, the International Radio Consultative Committee (CCIR), and the International Telegraph and Telephone Consultative Committee (CCITT). See, ITU, *International Telecommunication Convention, Nairobi, 1982*, Geneva, 1982, Art. 5.

7. For further information on the subject, see George A. Codding, Jr., and Anthony M. Rutkowski, *The International Telecommunication Union in a Changing World*, Dedham, MA: Artech House, Inc., 1982, Chapter 11.

8. See *Ibid*, Chapter 5.

9. See Radio Regulations, *op cit*., Art. 11.

10. See International Telecommunication Convention, *op. cit.*, para. 154. Emphasis added.

11. The ITU divides the world into three regions for the purpose of applying its Radio Regulations: Region 1,

which roughly encompasses Europe, Eurasia and Africa; Region 2, which includes the Americas; and Region 3, which comprises Asia and Australasia. See Radio Regulations, _op. cit._, Art. 8.

12. See U.S., Department of Commerce, _Report of the United States Delegation to the ITU Region 2 Administrative Radio Conference on the Broadcasting Satellite Service, Geneva, Switzerland, June 13 - July 17, 1983_, (mimeo.), pp. 2-3.

13. In October, 1983, the Federal Communication Commission stated that: "It seems patent that a detailed _a priori_ assignment plan contained in a treaty instrument for any communication satellite service other than broadcasting-satellite is not feasible, or to the extent feasible, not desirable. This planning approach utilizes the most inflexible possible legal device, a multilateral treaty, accepts the demands for allotments without any scrutiny or controls, explicitly or implicitly freezes dozens of technical and operational characteristics of every communication satellite facility at some low common denominator, impairs the ability to later share a common facility and suggests the conveyance of protection normally afforded an actual station. United States acceptance of such a plan would abrogate explicit policies maintained by

this Commission for more than thirteen years and be contrary to our nation's general policy on international radiocommunication agreements enunciated more than eighty years ago at the first multilateral radio conference." From U.S., Federal Communications Commission, An Inquiry Relating to Preparation for an International Telecommunication Union World Administrative Radio Conference on the Use of the Geostationary-Satellite Orbit and the Planning of the Space Services Utilizing It, Third Notice of Inquiry, October 7, 1984, p. 15.

14. See, T.V. Srirangan, "Equity in Orbit: Planned Use of a Unique Resource," paper presented at the IIC 1984 Annual Conference, Berlin, September 22, 1984, and China (People's Republic of), "Basic Considerations on Planning Objectives and Principles for the Space Service Utilizing the Geostationary-Satellite Orbit," CCIR, Preparatory Meeting ORB-85, Joint Meeting, Study Groups 1, 2, 4, 5, 7, 8, 9, 10, and 11, Geneva, 1984, Document B/43-E, May 14, 1984.

15. The Special Fund for Technical Cooperation was a voluntary technical assistance entity created by the ITU's 1973 Malaga-Torremolinos Plenipotentiary Conference over the objection of the United States. The United States has refused to contribute to the Fund

and most other developed countries have followed its lead.

16. See Codding and Rutkowski, *op. cit.*, p. 286.

17. The author also made this suggestion in an article prepared for the November 1984 issue of the *Harvard International Journal*.

18. See International Telecommunication Convention, *op. cit.*, Article 10 and 57.

19. For election purposes, the names of the candidates submitted by administrations are grouped according to the regions that they are from and all delegations are given the right to vote for one candidate in each regional grouping. For further information on IFRB elections, see Codding and Rutkowski, *op. cit.*, pp. 128-131.

20. See International Telecommunication Convention, *op. cit.*, Articles 10 and 58. With the exception of the U.S.S.R., most countries have been scrupulous in refraining from recalling their members during their terms of office.

DISCUSSION

QUESTION:

The ITU has been reluctant to accept U.S. contributions in the area of trained experts who might address LDC problems. This is based on some observations that the ITU is a European cartel. Do you believe that?

DR. CODDING:

The U.S. in in a phase in which it is cutting back on its contributions to LDC's and international organizations. There was an attempt after Nairobi to establish the Technical Training Institute. I don't know much about the Institute, but one Third World observer noted that it served primarily to advertise U.S. manufacturers' equipment. That doesn't seem to be the sort of thing I had in mind. There is a need for financial contributions. I don't think the ITU has ever refused contributions from any country.

QUESTION:

My point is that U.S. applicants to ITU positions have not been accepted for the most part. Europeans fare more favorably. Is this a valid observation?

MR. SRIRANGAN:

As far as the permanent headquarters staff is concerned, the U.S. is well represented at both senior and middle levels. The U.S. is not as well represented in the specific, shorter-term assignments. This may be attributed to the fact that these openings attract better qualified LDC candidates as a function of a differential wage scale which is not attractive for the really qualified U.S candidates.

DR. BLOCK:

Let me offer a few comments on the U.S. Telecommunications Training Institute. It arose out of an effort to more actively assist the developing nations regarding issues raised at the 1979 WARC and began a couple of years ago. In its first year, 200 LDC specialists came to the U.S. to receive training provided by ten of the major U.S. telecommunication companies. In some instances travel was provided by U.S. government agencies. This year the number of trainees has gone up to 450. There has been a very positive reaction to the program. It's an opportunity to receive state-of-the-art training, the sort of training the companies provide to their own employees. Despite the sales pitch, there appear to be benefits to be derived by all involved. There is now some discussion in the U.S. to send U.S. experts overseas for the same purposes with travel provided by the U.S. government.

DR. CODDING:

Is the U.S. Telecommunications Training Institute an academic institution?

MS. SANDRA LAUFFER, (Conference Participant):

It's administered out of Washington by the Academy for Educational Development for the board which is composed of U.S. Government representatives and presidents of major corporations. The training itself takes place at the corporate training centers.

QUESTION:

I'd like the panel to comment on Dr. Codding's proposals. How likely are they to succeed? What else might be considered regarding the capabilities of the IFRB?

MR. SRIRANGAN:

Proposals of this type have been raised before in the ITU. Two points are worth mentioning. In the first place, this issue cannot be examined at the 1985 conference given the established agenda. The conference in 1985 will address largely technical issues and will also evaluate in technical terms methodologies of planning and assist in the identification of frequency bands and space services where planning is required.

Anything that addresses regulatory matters to be modified or amended does not fall within the purview of the 1985 conference; it may be considered at the 1988 conference. So, there's little chance to discuss it in 1985, and it is too late to modify the terms of reference at the conference. It is a procedural problem.

The second point I wish to discuss is to what extent the mediating role is likely to be really useful. There is a provision written into the Radio Regulations now that any administration may seek the assistance of the IFRB in resolving problems in bilateral negotiations. But it cannot play the role of an arbitrator. If you confer upon the IFRB the powers of an arbitrator and the outcome is binding upon the participating countries, then it assumes the powers of the International Court of Justice. This, perhaps, was what they had in mind in 1947 at Atlantic City. But it brings with it its own host of problems. Decisions at such force take years to emerge. Since we are discussing the requirements of satellites in the near term and in an environment of rapid technological change, I have major doubts about such a procedure.

MR. TICE:

I don't think agencies of the U.S. government would submit to binding arrangements of this sort. If it were the least worst alternative, it might be considered. I'd like to ask Ed Probst if he has any thoughts on this.

DR. CODDING:

Point of information. I said mediation, not arbitration.

MR. PROBST, (Conference Participant):

I'd like to address the issue of whether or not the U.S. Government would be willing to make such contributions. I think it would be very difficult to do. Prior to 1973, the U.S. had contributed 55 shares of the ITU budget, 30 on behalf of the U.S. and 25 on behalf of U.S. territories. Modifications at the Plenipotentiary Conference eliminated the provision for membership of territories. Thus, the U.S. no longer contributed the 25 shares for territories. I argued within the U.S. Government that those funds should be added to its contribution to the UNDP or to direct assistance to the ITU. I have not seen change in the current Administration which alter its earlier response.

I'd like to pose a question to Mr. Srirangan. As you point out, the crux of the problem at the 1985 and 1988 Conferences has to do with the fixed satellite services. Given your assertion that regulatory procedures are inequitable, can you foresee any changes in regulatory procedures dealing with the other 16 or 17 satellite services?

MR. SRIRANGAN:

The issue won't be addressed formally until 1988. If it is decided at the 1985 conference that these services can be addressed collectively within a regulatory framework, and not a planned framework, then a regulatory mechanism can be defined in greater detail for the 1988 conference. Guaranteed and equitable access should permeate all provisions. There should be a clear set of guidelines.